Russell Thacher Trall

Diphtheria

Its nature, history, causes, prevention, and treatment on hygienic principles; with a

resumé of the various theories and practices of the medical profession

Russell Thacher Trall

Diphtheria
Its nature, history, causes, prevention, and treatment on hygienic principles; with a resumé of the various theories and practices of the medical profession

ISBN/EAN: 9783337311834

Printed in Europe, USA, Canada, Australia, Japan

Cover: Foto ©berggeist007 / pixelio.de

More available books at **www.hansebooks.com**

DIPTHERIA:

ITS

NATURE, HISTORY, CAUSES, PREVENTION,

AND

Treatment on Hygienic Principles;

WITH A

RESUMÉ OF THE VARIOUS THEORIES AND PRACTICES OF THE MEDICAL PROFESSION.

BY

R. T. TRALL, M.D.,

AUTHOR OF THE "HYDROPATHIC ENCYCLOPEDIA," AND OTHER WORKS; PRINCIPAL OF THE HYGEIO-THERAPEUTIC COLLEGE; PHYSICIAN-IN-CHIEF TO THE NEW YORK HYDROPATHIC AND HYGIENIC INSTITUTE, ETC., ETC., ETC.

NEW YORK:
FOWLER AND WELLS, PUBLISHERS,
No. 308 BROADWAY.
1862.

Entered, according to Act of Congress, in the year 1862, by

R. T. TRALL,

In the Clerk's Office of the District Court of the United States for the Southern District of New York.

PREFACE.

The increasing prevalence of the malady known as diptheria, in various parts of the United States, the disastrous results of drug-medication, and the superior safety and efficacy of the Hygienic or Hygeio-Therapeutic method of treatment, supply the motive for presenting the public with a monograph on the subject. During the last two years I have been written to for information by hundreds of heads of families, in neighborhoods where two, three, four, five, six, seven, eight, and in one case *nine* members of the same family have died of this disease, or of the treatment, or of both combined; and everywhere the physicians seem to be in doubt as to its real pathology or proper treatment, while the people are in consternation because of its direful ravages. It has been my fortune to see much of the disease, and to have been in correspondence with many of the graduates of my school, as well as many other professional and non-professional persons who have successfully applied the plan of treatment recommended in this work. And these circumstances have induced me to collate the substance of nearly all that has been published on the subject in this country and in Europe, and to note the facts and statistical data which have been presented to the profession and the public through the medium of the journals of the different medical schools. The work, therefore, here offered to the public is intended not only as an exposition of the true pathology and proper management of diptheria, but as a record of all that is important which has been ascertained in relation to the disease and its treatment to this date.

R. T. T.

New York Hygienic Institute, No. 15 Laight Street.

CONTENTS.

	PAGE
Nosology and Technology of Diptheria	7
Description of Diptheria	10
Pathology of Diptheria	38
The False Membrane	65
History of Diptheria	76
Infectiousness	87
Causes of Diptheria	95
Mortality of Diptheria	103
Complications	116
Sequelæ of Diptheria	120
Morbid Anatomy of Diptheria	134
Drug Treatment of Diptheria	158
Hygienic Treatment of Diptheria	227
Tracheotomy	258
Stimulation vs. Antiphlogistication	261

DIPTHERIA.

NOSOLOGY AND TECHNOLOGY.

THE malady now generally known as *diptheria, diptherite,* or *diptheritis,* has been a perplexing theme to the nosologists ever since its first recognition as a distinct disease. Some have classified it with *croup;* others have regarded it as a modification of *malignant scarlatina,* while others have considered it to be an affection *sui generis.* It certainly presents many of the characteristics of that form of scarlet fever which has prevailed extensively in many places under the name of *putrid sore throat,* while the exudation of a fibrinous material on the mucous surfaces seems to ally it closely with croup.

The term *diptheritis,* in its most literal signification, implies *inflammation of the skin.* By some authors it is employed in a generic sense, to include all affections of the mucous membrane characterized by the exudation of any material capable of coagulating or concreting into a membranous covering or coating. In this sense it comprehends the *true croup,* this being a variety or species of diptheritic inflammation.

Professor George B. Wood, M.D., of Philadelphia (Wood's Practice of Medicine), denominates it *pseudomembranous inflammation of the fauces.* The term *membranous quinsy—angina membranacea—*has been applied to it by some authors. *Stomatitis* and *follicular inflammation* are also among the numerous technicalities with which a medical literature, more wordy than wise, has served to confuse and mystify the subject.

M. Bretonneau, who first explained the precise nature of the excretion of the disease, applied the term diptheritis to a group or class of diseases which affect the dermoid tissue—the skin and mucous membranes—and which are characterized by a tendency to the formation of false membranes. In this sense, the term would be applicable to certain cases of *catarrh in the bladder*, *tubular diarrhea*, and *dysmenorrhea*, for, in all of these cases, there is frequently a similar exudation of lymph, and the formation of preternatural membranous concretions, which are cast off in fragments, and often accompanied, especially in the case of the uterine affection, with excruciating suffering. These morbid membranous formations have not unfrequently been mistaken for a sloughing or casting off and discharge of the mucous membrane itself.

A malady closely allied, if not identical with diptheria, or *scarlatina maligna*, has prevailed as an epidemic among domestic animals as well as among human beings. Thirty or forty years ago it prevailed extensively in western New York, and was there termed "*black tongue*." It was very fatal, as is the disease of human beings known as "putrid sore throat."

"The medical history of the present century," says Edward Headlaw Greenhow, M.D., of St. Thomas' Hospital, London, in a late work on diptheria, "is remarkable for the reappearance in this country of two very definite forms of epidemic disease described by the physicians of former centuries, but unknown to our immediate predecessors. I have elsewhere shown that the disease which in our day is called Asiatic or epidemic cholera, is identical with a disease named *dysenteria incruenta* by Willis, and *diarrhea colliquativa* by Morton, which prevailed during many years of the middle and latter part of the seventeenth century. The kind of epidemic sore throat, now called diptheria, which has prevailed so extensively during the last four years, though unknown to the last two or three generations of physicians, was familiar to the medical practitioners of this country about the middle of the eighteenth century, under the names of malignant sore throat, epidemic croup, and *morbus strangulatorius*. Both cholera and diptheria have, it is true, been observed from time to time in a sporadic form; and small outbreaks of each of these diseases have sometimes occurred—but in an epidemic form they had been long unknown, when they reappeared in our own time."

"The terms cholera and diptheria are, generally speaking, and perhaps properly, only applied to the malignant forms of these epidemic diseases, to the exclusion of the milder and commonly more numerous cases of illness induced by the epidemic influence. These milder cases, although characterized by an affection of the same mucous surfaces, lack the more striking features usually understood to be associated with the terms cholera and diptheria. The mucous surface of the alimentary canal is alike the seat of the principal phenomena, both in cholera and the diarrhea which commonly prevails so extensively during a visitation of cholera. The mucous membrane of the throat, especially of the tonsils and immediately adjacent parts, is not only the seat of the simpler form of sore throat which has prevailed so extensively during the last three or four years, but is likewise, almost

1*

invariably, the situation in which the first symptoms of the more severe cases, properly termed diptheria, manifest themselves. The diarrhea of cholera times does not present the excessive prostration, the blue, cold, clammy surface, the pulseless extremities, or the whispering voice of fully developed cholera; the simpler sore throats which have usually prevailed simultaneously with diptheria have been often unattended by the characteristic exudation of false membrane, or by the prostration of strength, and have rarely, if ever, been followed by the rancous nasal voice, the paralysis of the muscles of deglutition or of locomotion, and the impaired vision which so frequently follow in the train of diptheria; but the diarrhea and sore throat are respectively congeners of cholera and diptheria, from which their difference is less one of character than of degree."

Dr. Greenhow limits the term diptheria to the inflammation of mucous surfaces characterized by membranous exudation, but employs it generically to comprehend all forms of sore-throat affection attended with this exudation.

DESCRIPTION OF DIPTHERIA.

The following description of the disease, taken substantially from the late work of Dr. Greenhow, is the most accurate and carefully drawn of any which I have seen, as applicable to the great majority of cases:

"Diptheria, comparatively rare as a sporadic disease, prevails as an epidemic, in which form it often exists cotemporaneously over considerable tracts of country, or it may occur in smaller groups, limited to particular hamlets, or even to particular houses. Sometimes it has prevailed so extensively that distant countries, including portions both of the Old and New World, have been simultaneously or successively visited by it.

"Diptheria is sometimes preceded, and usually accompanied, with fever, which, in certain epidemics and in severe cases, is only transient, speedily giving place to depression. There is often a stiffness of the neck at the commencement of an attack, and usually more or less swelling and tenderness of the glands at the angles of the lower jaw. The tonsils are commonly swollen, and, together with the immediately contiguous parts of the mucous surface, more or less inflamed. Sometimes the swelling and inflammation subside without further local mischief; at others, the inflamed surface presents, from an early stage of the disease, whitish specks, or patches, or a continuous covering of a membraniform aspect, which may appear as a mere thin, almost transparent pellicle, but usually soon becomes opaque, and in some cases assumes the appearance of wet parchment or chamois leather. This membranous concretion varies in color from being slightly opaque to white, ash-color, buff, or brownish, and in rarer instances to a blackish tint.

"This false membrane is a true exudation which has coagulated upon the mucous surface, from which it may often be readily separated, leaving the subjacent membrane mostly unbroken, or merely excoriated, usually reddened, vascular, tender, and dotted with small bloody specks or points, but sometimes superficially ulcerated, and more rarely in a sloughing condition. When the false membrane has been artificially removed, it is apt to be renewed; and when not meddled with, to become thicker by continued exudation from the mucous surface. The severity of the disease is commonly in proportion to the continuity and density of the exudation; but cases sometimes occur in which the membranous exudation is inconsiderable, and yet the general symptoms are of a very alarming kind. If the patches are small and remain distinct, the case ordinarily runs a favorable course; if they rapidly spread and coalesce, if the membrane becomes thick, and especially if it assumes a brownish or blackish color, danger is imminent. In proportion as the membrane increases in thickness and density, does its attachment to the subjacent surface generally become firmer. The surface of the mucous membrane around the exudation is red and vascular, and so tender that in severe cases it bleeds on the slightest touch.

"The throat is in general the primary seat of the disease; but the inflammation is apt to spread along continuous mucous surfaces, and thus to extend upward into the nares and to the conjunctiva; down the pharynx into the esophagus; through the glottis into the larynx, trachea, and downward into the bronchial tubes; or forward on to the buccal mucous membrane, the gums, and lips. Wounds and excoriations of the skin, and the mucous membrane of the nymphæ and vagina when tender or irritated, especially in persons already suffering of diptheria of the throat, are during an epidemic liable to undergo the same process of exudation, which, coagulating, forms a false membrane analogous to that on the tonsils and throat.

"Albuminaria, commencing early in the disease, usually within a few hours, and gradually disappearing with the local affection, sometimes, but by no means invariably, accompanies diptheria. If the urine be much loaded with albumen, the complication is a serious one; but cases have done well in which a considerable cloud of albumen was deposited from the urine by the proper tests, and very severe and fatal cases of diptheria have been unattended with albuminaria.

"After a time the false membrane is thrown off, either entire, so as to represent a mold of the parts it covered, or, which is more usual, comes away in shreds or flakes, intermingled with mucus. Sometimes it undergoes decomposition prior to separation, giving rise to a very offensive smell. When the membraniform exudation has come away spontaneously, it is sometimes repeatedly renewed, each successive false membrane becoming less and less dense, having less and less of the character of exudation, and more and more of that mucous secretion, until at length the affected surface is merely covered with a thick mucus, which gradually disappears as the mucous membrane recovers its healthy condition. In other cases the exudation is not renewed when it has once been thrown off, but the subjacent membrane is observed to be either redder or paler than natural, has a rough, rugged appearance, or is depressed below the adjacent surface on the parts where dense false membrane has existed. Occasionally sloughing takes place beneath the exudation, or even more deeply, as in the center of a tonsil, and may implicate the tonsils, uvula, and soft palate. More rarely the tonsils suppurate. Hemorrhage from the nose and throat, independently of the co-existence of purpura, often occurs in the course of diptheria, and is sometimes very profuse. The local affection may pass into a chronic form, in which relapses or exacerbations are readily produced by vicissitudes of weather, or by exposure to damp or cold. Even perfect recovery from an attack affords no immunity from the disease in future.

"A peculiar character of the voice, resembling that produced by affections of the throat in secondary syphilis, is a common result of diptheria, and often continues for many weeks after recovery. The power of swallowing is sometimes so impaired that there has been difficulty in sustaining life during convalescence; and the liquids especially are apt, even after a comparatively slight attack of the disease, to be regurgitated through the nostrils. Extreme anemia, impairment of vision, a peculiar form of paraplegia, weakness of the hands and arms, numbness, tenderness of the limbs, tingling, wandering pains, and more rarely, nervous sequelæ of a hemiplegic character, are, in the order here written, ulterior consequences of diptheria. Gastrodynia, and sometimes dysenteric diarrhea, occasionally follow diptheria. Pain of the ear, deafness, and abscess are occasional but rare results of the disease."

In *Braithwaite's Retrospect* for January, 1860, is a graphic and very accurate description of the disease in its severer form, by James P. McDonald, Esq., of Bristol, England, who has treated a large number of cases:

"I consider diptheria to be a disease produced by a specific poison taken into the system, acting through the blood and *seen* at the throat."

In the above passage the author confounds the poison itself, or *cause* of the disease, with the *effect*. What is seen at the throat is not the specific poison which induces the disease, but the excretion by means of which the living system undertakes to cleanse itself of the morbific material, and the idea that the poison "acts through the blood" is a part of the false theory of disease entertained by the whole medical profession.

Poisons do not act on nor through the blood, nor any of the tissues or organs. They do not *act* at all. But, on the contrary, the living machinery *acts* to expel them, *and this expulsive process is the disease*, strange as this announcement may sound to ears unaccustomed to regard disease as a process of purification and reparation—a remedial effort.

Mr. McDonald continues: "The following are the usual form and course of the disease in its severest type. The patient is *suddenly* (and generally in the morning) seized with violent vomiting of a thin, yellowish-white matter of a very offensive character; then purging of a fluid of similar appearance and smell. These dejections last an hour or so, and are followed by great prostration and stupor. The patient lies for a period varying from six to sixteen hours in a heavy sleep, from which he is with difficulty aroused, and then only to sleep again. The skin is hot; pulse 100; the tongue is of a bright red; drink is taken with avidity, if offered, but only to be immediately returned. And now the important question is put, 'Is the throat sore?' The answer is *always* the same—'*not in the least.*' The reply, to the inexperienced in the horrible malady, may be fatal to the patient. The diagnosis is that this is not a case of diptheria. On the other hand, the experienced man *expects* this reply; he forthwith carefully examines the throat, and then he *sees* the disease. In this early stage the tonsils, the soft palate, and the back of the pharynx present a bright, shining red appearance. The small vessels are not seen individually injected, as in many forms of sore throat, but the appearance is as though the parts had been brightly painted and then varnished. Hanging from the velum to the tongue is seen, in this stage, a transparent film of a tenacious fluid, which is burst by expiration, sending its particles over the mouth and the instrument used to depress the tongue. The next moment a similar curtain is formed. After a period varying from six to sixteen hours, the condition of the patient materially changes. The stupor has passed off, and delirium, often of a violent character, takes its place; there are the usual symptoms of cerebral excitement, and the fever runs high; breathing is quickened; the voice is changed to a thick yet shrill tone; there is a short, dry cough (in children evidence of coming croup); the neck is puffy and blushed; the tongue is coated with a white fur, and all those parts hitherto so brilliantly red, are thickly spotted with a whitish substance, which, in a wonderfully short period, conglomerates and forms one thick, plastic deposit, which in time may cover the whole palate to the teeth, so that the appearance, on opening the mouth, is as though it were lined with plaster-of-Paris. The violent delirium then subsides; the powers of life fail rapidly; the horrible sensations of choking and suffocation come on; the sufferer tears at his neck with his nails, and tries to open his mouth, yet full power of swallowing still continues, and he greedily gulps anything given him in the shape of drink; large livid spots form on the extremities, amounting sometimes to purpura; the diarrhea, of a white and offensive matter, is increased; muttering delirium comes on, and in a long tetanic convulsion, death closes the scene.

"This is a truthful picture, drawn from realities, of how a previously strong and healthy man may, in six days or less, cease to be.

"Taking the above as a fair example of diptheria in its most marked and deadly aspect, as I have seen it, we get the resemblance to it, more or less, in all minor cases. We must not expect to meet with all the symptoms in every case, but the condition of the throat is invariable. Whether that condition goes on to the second stage, depends on the severity [quantity?] of the poison or the success of the treatment adopted. In all cases where there is either nausea or vomiting, followed by drowsiness, the throat ought to be examined, and if the redness and the 'glassy curtain' appear, the immediate use of the proper appliances may, I am quite certain, save many valuable lives."

In the same journal is an article from the pen of Thomas Heckstall Smith, Esq., surgeon to St. Mary Cray, Kent, who distinguishes three forms of diptheria: "There are three forms in which the disease presents itself, viz., simple ash-colored diptheria membrane in patches, with very slight congestion of the surrounding parts, and without fetor. Secondly, a deeper color, and more widely-spread membranous exudation, with fetid breath and intense engorgement of dark hue. Thirdly, the membrane with much tonsillitis, in a few cases resulting in quinsy. But there has been a fourth and more formidable state of things to contend with, namely, an extension of the membrane in either of the above forms, to the larynx and trachea, the symptoms of which I need not describe."

In a paper read before the New York Academy of Medicine, January, 1861, by David Winne, M.D., of the University Medical School, the symptoms are thus described: "Diptheria is frequently attended with very slight constitutional disturbance at the commencement of the attack, even where the disease is destined to a fatal termination. The patient is often so little affected, that, with the exception of some slight difficulty in the act of deglutition, he exhibits no evidences of disease, and it is with difficulty that the parents can be brought to consider this symptom as one of much importance, or the child in very serious danger.

"After a short interval, however, one of the tonsils—seldom both—becomes specked with a yellowish-white deposit, which, when seen at this early stage, presents the appearance of small whitish stars in the midst of a ground of what appears to be a transparent layer of mucus, but which really is the true diptherial membrane through which the body of the tonsil, often of an increased redness, is distinctly seen. These spots, small at first, rapidly enlarge, the membrane loses its transparency, and if not speedily arrested, spreads over the soft parts of the palate, both tonsils, the uvula, and involves the larynx, and sometimes the trachea and bronchial tubes.

"Usually, even in slight cases, the local symptoms are preceded by some constitutional disturbance. There is a feeling of malaise, pain in the head, often extending to the neck; lassitude, and more or less fever. In the mild form the tongue presents a thick creamy coat, through which a few papillæ are visible; the uvula, the velum palati, and pharynx are of a bright red color, and the tonsils swollen and specked with a filmy deposit, already described, which is generally closely adherent to the mucous membrane, although in some cases it is easily removed in its earlier stages by the application of the sponge probang, which is often coated with the new-formed deposits.

"This membranous exudation may extend over the whole palate, but in mild cases rarely does; nor is its color much deepened, or the odor emitted offensive or fetid. The submaxillary glands are slightly swollen, but do not attain the size which they acquire in the severer forms of the disease. Under favorable circumstances, or the application of judicious treatment, its progress is here arrested. The membrane ceases to spread, and slowly becomes detached from its connections; the mucous membrane loses its red color; the glandular swellings subside; the pulse diminishes in frequency, and the patient becomes decidedly convalescent.

"The disease, however, does not always present itself in this form, but is ushered in by rigors and often vomiting, under whose influence the patient becomes so prostrated, that it soon becomes obvious that the system is oppressed by a powerful poison. This condition is characterized by a high [violent?] fever, a pungent skin, a rapid and feeble pulse, great difficulty in deglutition, hurried respiration, flushed countenance, and congested lips; the tongue becomes loaded with a yellow or dirty brown coat; the soft palate and pharynx assume a deep erysipelatous redness; the tonsils become greatly swollen, and the ash-colored membrane, nearly continuous and spread over one or both tonsils, extends to the uvula and the posterior walls of the pharynx. As the disease advances, these symptoms increase in severity; the breathing becomes more hurried and stertorous; the swallowing, which at first was but moderately impeded, becomes so troublesome and painful, that the child is with great difficulty induced to take either food or medicine; the saliva flows from the mouth, and often a foul and acrid discharge from the nostrils. Should the little patient be induced to swallow, food or drink will be violently ejected, and a paroxysm of great intensity, in which the child will gasp for breath, and with great difficulty recover itself, will ensue.

"The case has now reached a point which portends the most unfavorable results. The false membrane has seized upon every visible part of palate and pharynx; the discharge of sanies mixed with blood, which issues from the mouth and nose, has become exceedingly offensive; the glands of the neck become enlarged and tender, the voice hoarse and indistinct, the pulse more rapid and feeble, and the poor patient, restless and embarrassed for want of breath, tosses about or lies on his back in a semi-comatose state; in most cases the medical attendant is apprised by a croupy respiration when the membrane has invaded the larynx and trachea, at which time symptoms of asphyxia present themselves; the countenance becomes livid, the skin cold, the pulse feeble or gone, and the patient, either distressed for want of breath, anxiously awaits the moment when death shall relieve him of his sufferings, or rapidly sinks into an asthenic or comatose condition."

The careful reader will not fail to notice some discrepancies in the symptoms, as described by the various authors thus far and hereafter to be quoted, a fact which shows the great diversity of forms and the various degrees of malignancy under which the disease appears in different persons.

A. C. Hamlin, M.D., Surgeon 2d Regiment Maine Volunteers, in an article published in the New York *Medical Times* of Feb. 22, 1862, makes the following remarks in relation to diptheria as it prevailed in his department of the army: "Since the commencement of the campaign, some thirty cases of diptheria have been observed by us, most of which have been so obscure and complicated as to render diagnosis perplexing, and often inclining us to doubt whether the malady merited a distinction from some other phlegmasias of the throat by reason of functional symptoms and physical signs. Rarely did it commence with the pellicle of Bretonneau, though it afterward assumed many of the peculiarities of the disease in an advanced stage. Sometimes the exudation appeared like cryptogamous vegetation; then, again, there were ulcerated fissures or irregular patches with flake-like lymph. All the cases appeared during or after wet and stormy periods, when the atmospheric variations were sudden and the electric oscillations considerable. All ended in resolution, without serious injury to the system except one, in which instance death ensued from hemorrhage of the palatine or pharyngeal arteries. The enlargement of the cervical glands was often very great, with occasional abscess. The attending pyrexia [fever] and constitutional disturbance were in most cases slight."

Dr. Fouregaspd, of Sacramento, California, thus describes diptheria as it appeared in that place:

"The disease begins in a very insidious manner, by a little engorgement or inflammation of the soft palate, pharynx, and one of the tonsils. At this period the patient complains but little—there is no fever, or it is very moderate. The pain in the throat is much slighter than in the usual forms of sore throat—so slight, that the little patients go about playing as if nothing was the matter. In some exceptional cases, the fever and inflammation about the pharynx are considerable from the beginning. The characteristic signs of the invasion soon follow. They consist in small portions of white or yellowish lymph deposited on the palate, the tonsils, and the posterior part of the pharynx. The cervical and submaxillary glands become swollen, and the pain in swallowing and opening the mouth is occasioned more by the engorged state of the glands than by the internal secretion of lymph. These deposits go on increasing in size more or less rapidly, and in violent cases in a few hours the whole cavity of the throat is covered by them. Generally one side is more affected than the other, and the glands corresponding with the parts affected are more swollen than those of the opposite side."

Dr. Blake, of California, in the *Pacific Medical and Surgical Journal*, August, 1858, describes the *access* and progress of the disease: "Drowsiness, prostration, or oppression is manifested by infants, or complained of by adults; and when the disease is prevailing, this desire of children to sleep at other than usual hours should awaken our suspicion. The pulse is accelerated from the first, but generally soft and typhoid; although in some cases it is for a few hours rather hard. The temperature of the skin is raised, although seldom harsh or dry; and frequently moist, or even covered with profuse perspiration. There is seldom any pain; rarely headache or backache. The tongue is usually coated, edges red, papillæ prominent. The appetite may remain good, and the digestion unimpaired. If we examine the throat we may find, even within twelve hours after the occurrence of the first slight symptom, the tonsils covered with a gray, pultaceous exudation, which rapidly extends upward into the nostrils, and downward toward the larynx; and again, we may detect only a redness of the tonsils and a small point of exudation, two or three days after the commencement of the disease, and at a time when the symptoms of general prostration had become alarming. Again, cases may present themselves in which the general symptoms and the anatomical lesions proceed *pari passu;* but in almost every case that I have seen I have considered that death was the result rather of the action of the poison on the system than from obstruction of the larynx. In from twelve to twenty-four hours after the formation of the false membrane, we generally find the cervical glands enlarged, and in severe cases this enlargement may afford a serious obstacle to respiration and deglutition."

Professor Alonzo Clark, M.D., of the New York College of Physicians and Surgeons, in his lectures on diptheria, as published in the New York *Medical Times,* distinguishes between epidemic sore throat, which has little tendency to the production of membrane, and true diptheria—the former being comparatively a mild disease. The difference, however, may be more in degree than in kind.

Dr. Clark, limiting the term, diptheria, to such forms of inflammation as terminate, or have in their course this membrane as a sign, thus describes the order of symptoms:

"As to its initiatory symptoms, they have no definite relation to the future severity of the disease or to the parts that are to be the seat of the inflammatory exudation. When diptheria appeared among us for the first time as a prevailing disease, the cases that I saw were almost all of them ushered in by pretty acute symptoms; a chill, followed by a fever; and then, in a small proportion of cases, a chill and fever alternating two or three times in the course of a single day. Those instances in which the chill was repeated were rare; but a very decided invasion was, in the cases that I saw, the rule in the beginning of the disease. As it went on, the symptoms of invasion were less and less marked, and not unfrequently, as is now noticed, it occurs without any that attracted attention. Several instances of this kind now occur to my mind; but two of these will serve for illustration:

"Two children, two and a half and four years of age, were observed to have the symptoms of slight catarrh for two or three days, but there was nothing to awaken anxiety. They followed their amusements in the nursery as usual, when at length the mother noticed a croupy cough in the youngest, and sent for the family physician. He found the usual early symptoms of croup, and a diptheritic membrane on the tonsils, extending downward beyond the reach of sight. He examined the other child's throat, not because he expected to find any evidence of grave disease, but from motives of prudence, and was surprised to find the tonsils almost completely covered with false membrane. The youngest grew rapidly worse, and in four days died of diptheritic croup. The eldest was at no time dangerously sick, and did not keep her bed a single day. The membrane was detached in two days, and did not reappear. The only medicines were tonics and chlorate of potassa, with full nutrition. Bretonneau, in examining the throats of young persons in a school where diptheria was prevailing, found the membrane in many instances where there was no complaint of ill health, and where it was not suspected till it was actually found. Such cases will teach you two important lessons: first, that the disease does not always make its invasion by any symptoms calculated to excite alarm; and secondly, that those symptoms, when once declared, are to be considered by no means as a measure of its severity. It is not easy, then, to fix in very definite terms the character of the invasion, the symptoms being sometimes very decided, at other times very insidious. But where the disease is once formed, you look for symptoms relating to the fauces, trachea, nasal passages, mouth, or esophagus, for it is in these that the membrane is most frequently formed.

"When it is confined to the *fauces*, there is often but little occasion for alarm. These are the cases from which most of the recoveries come. The breathing is not interfered with; there is not necessarily much cough; the general health may not suffer materially. And yet, let me say to you, that when it forms in the fauces only, and does not extend beyond, you will not unfrequently find, as the disease advances, the most formidable symptoms; and as we shall see, by-and-by, too often a fatal result.

"When it advances into the *nasal passages*, you will have indications somewhat before the formation of the membrane. You will usually see it in the fauces, perhaps folding back beyond your view upon the palate; the nose will become a little red, and there will be a little snuffling upon one or both sides; directly there is a discharge of a yellowish watery or ichorous matter, nearly transparent. This may irritate the skin of the lip a little, and may, in the end, cause swelling of the upper lip itself. Soon after this discharge makes its appearance, there may be seen forming upon the swollen mucous surfaces a delicate membrane, and this, growing thicker and more abundant, will not unfrequently stand out upon the white tissues joining the red of the nose. And then still the ichorous matter will continue to be discharged; it will sometimes dry up on the false membrane, and finally plug up the nostrils altogether, so that respiration can be performed only through the mouth. At other times the nostrils are not plugged up, and breathing through them is only difficult.

"When the membrane forms in the *esophagus*, you have no very decided indications of its presence there. There is no great difficulty of swallowing; there is no particular pain that will lead you to the suspicion of its formation in that tube. You learn it mainly from the fact that ribbons, or a large membrane, are vomited up, or perhaps the same things may be found in the stools.

"But when the *larynx and trachea* are invaded, you have the most formidable variety of this disease. Then it is that you have everything to fear. Then the chances for recovery are scarcely so good as one in eight or ten of all who are attacked. The symptoms of this invasion of the trachea and larynx are precisely or almost precisely those of croup. The voice is changed; it loses its compass and strength, and frequently is reduced to a whisper. The breathing becomes noisy; we call it stridulous; the cough, for the most part, becomes hoarse and croupy—occasionally shrill and brassy; there is difficulty of breathing; the child's head is thrown back to open the larynx fully and give force to some of the respiratory muscles. He not unfrequently vomits, but this affords him very little relief. The difficulty of breathing becomes more and more considerable as the disease increases, and in some instances there is very marked restlessness. In other instances there is much drowsiness. The surface of the body often shows the marks of incomplete aeration of the blood. The nails and lips become blue, or there may be a general cyanotic condition. The wings of the nose are expanded in inspiration. Everything shows that the child is about to die from asphyxia or apnœa. While in the other forms of the disease children die from the general influences of the diptheritic poison, these scarcely live long enough to experience them.

"This membrane may be found *lining the whole mouth*. Then it usually is produced first in the fauces, and extends forward. In my observation, the mouth does not take on this diseased action in the mild cases, but rather in those in which the disease is invading the nasal and respiratory passages. It has been known to begin on the gums (gingival diptheria), and extend backward into the fauces, so covering the mucous surfaces of the mouth. When the mouth is so covered, the red tissues are everywhere—on the roof, the inner surface of the cheeks, the tongue, the gums—hidden by a layer of exudation that looks like a half-dried coating of plaster-of-Paris. As this peels off portion by portion, the natural structures are left red and shining. This stomatic diptheria alone is no more grave than other forms of the same disease, and much less so than the tracheal variety. It produces but little of actual pain, but it makes the mouth stiff and embarrasses its motions; destroys the taste for the time; makes it painful to talk and swallow. Hot and stimulating drinks appear to be in the highest degree unpleasant. Indeed, the little sufferers affected in this way sometimes resist every administration by the mouth with a perseverance—I may even say a frenzy—which only an absolute and apparently cruel firmness on the part of attendants can overcome.

"In all these forms of disease one feature is almost uniformly noticeable, and that is a *swelling of the glands* at the angle of the jaw, and of those extending downward from this point. Indeed, it is regarded as one of the diagnostic marks in the early stage that these glands, though ever so little, are swollen. They are usually swollen *unequally*. When the disease is prevailing, Bretonneau warns us, at the least snuffling, on the slightest indication of coryza, to feel behind the angle of the jaw, and below the lobe of the ear, and so down the side of the neck for swollen lymphatic glands. We are then to examine the upper lip. 'In simple coryza the skin is reddened equally under each nostril, while in the Egyptian disease it is only on the side of the glandular swelling. If the swelling exists on both sides, it is unequal. On the side where the swelling is least, the redness of the lip will be least. From the period of this discovery we are certain there is a special affection—in fact, the Egyptian disease.' By 'Egyptian disease,' M. Bretonneau means diptheria.

"In this connection, I may better say that this disease may appear *on the gums*, as it often appears on the tonsils, without extending beyond the parts it first attacks. Such cases belong, in general, to the milder forms of diptheria.

"Among the rarer seats of diptheritic exudation, I may mention the *external ear*. This tube has been seen lined by it. M. Bretonneau reports an instance in which *the lining membrane of the antrum highmorianum* was fully involved. A poor Jew had died while the physician was making preparations for tracheotomy. The false membrane was found in all the air passages as far as they could be followed, and also making *an adventitious lining of both maxillary sinuses*, filling both with a turbid serous fluid, in which were floating bands of false membrane, as in a pleuritic effusion.

"I have here a letter from Dr. Whittlesey, physician

to the Children's Hospital on Randall's Island, relating to some cases of *diptheritic ophthalmia* that occurred there some time ago. Dr. Rives, assistant physician in that institution, two or three years ago, exhibited to me some specimens of this disease, and they were shown to the class then attending lectures here. The eyelids were both covered by a firm, elastic exudation, and the same membrane covered the conjunctiva of the eye as far as the cornea. Dr. Rives informed me that in his department of the hospital there had been at that time five cases of this affection, more or less extensive, and that in his cases, if the patient survived, the inflammation was destructive to the eye, and blindness followed. Dr. Whittlesey's letter informs me that these cases occurred in the winter of 1857 and '8, before diptheria became epidemic in this city, and while it was prevailing in Albany. But a similar disease showed itself in that institution four years earlier. Dr. Whittlesey states that, 'In the winter of 1853–4 measles and scarlet fever prevailed in this institution, and there were three cases of diptheria. The patients were children that had suffered from measles, and were in a feeble, emaciated condition. They all died in a few days after the membranous disease appeared. The deposit or exudation was upon the inside of both eyelids, nearly a line in thickness on the upper, and of such consistence that it could be removed with forceps, retaining the form of the lid as a cast, presenting an appearance similar to that of the specimen presented to you by Dr. Rives.' This form of diptheria has been repeatedly noticed in Europe.

"These are, however, only the local manifestations. Those of a more general character are still to be considered. It not unfrequently happens that persons who have gone through with all that I have now described to you, and appear to be recovering, suffer still from a prostration that seems almost unaccountable. Take one or two fatal examples. Early in the occurrence of the epidemic, in a patient of Dr. Crane's, the membrane, if I remember rightly, was found, as it is commonly, in the fauces, but not beyond. The patient went through with the earlier stages of the disease, the membrane exfoliated, and everything seemed to be doing well. His convalescence was announced to the friends of the family. About ten days after the membrane disappeared, Dr. Crane was called in haste to see the child, as it was very much worse. When he reached the house, he found that he was so much prostrated that there was scarcely any pulse. The patient had been sitting up the earlier part of the day, but now he could not raise his head from the pillow without fainting. It seemed to the Doctor that there was internal hemorrhage, yet there was no other manifestation of it. In this sinking condition the little one remained from two in the afternoon until seven in the evening, when he died, precisely, if I can judge, as persons usually die from the rupture of some vessel that allows fatal hemorrhage into the intestines or uterus. On the morning of the day on which he died, there was nothing to lead to the suspicion that he would not get well, éxcept the treacherous nature of the disease. In Dr. McCready's case, already referred to, a similar history is to be given. This child had an extraordinarily thick membrane formed upon the tonsils and uvula; you see a portion of it in that vial. The symptoms were those of ordinary sore throat at first. In a day or two the tonsils became covered with the membrane. There was not much disturbance of the general health. In a few days exfoliation took place, and there was promise of speedy recovery. A week later, however, membrane appeared in nostrils; rapid collapse followed, and the child died in twenty-four hours.

"A son of Mr. D., two years old, had the diptheritic membrane first in the fauces, afterward in the larynx, and probably in the trachea. Little hope was entertained of his recovery for many days. At length the croupous cough, the rapid and stridulous breathing slowly subsided, with the expectoration of fragments of membranous matter, and the child appeared to be convalescent. The danger seemed to have passed, and he was taken into the country. But there he lost strength and flesh, sank into deep prostration, and died in three weeks without renewal of the dyspnœa, or any other symptom of throat disease.

"Well, now, what is it that produced death under these circumstances? The obvious answer is—a certain poison, the nature of which we do not understand, which, though it has spent its force to produce local manifestations, has not yet exhausted its fatal control over the nervous system. It seems to destroy, making allowance for the difference in time, as prussic acid does, by overwhelming the nervous forces. I know nothing else to say about it. A case or two more to illustrate this point. In a patient on Staten Island, whom I saw with Dr. Gunn, the history is a little different, and yet no more favorable. A young lady, fourteen years of age, had the membranous disease of the fauces; it was of the variety once called the sloughing sore throat. A membrane had formed of considerable firmness and thickness, and apparently in successive layers; the older parts were sloughing off from the newer. Her throat looked as if there was an abundant dirty purulent slough covering it. This is no uncommon appearance; and these very appearances have led to some of the names which have been given to this membrane in the older time. You can hardly believe when you see such an appearance that it is not really a gangrenous condition of the natural tissues of the parts; but if you watch such a case, and it has a favorable termination, you will see that the whole of this material will clear off without even so much as a depression being left. This was the condition of the young lady's throat. Her breath was somewhat, but not markedly fetid. She had been sick just six days, when I saw her. She had been attacked with sore throat pretty suddenly in church. Not having a chilly feeling, but still experiencing general discomfort, she left the church for her father's house. The physician was called the next day, and found the membrane. It continued then, from Monday until Saturday; and now, without any great loss of strength, without any difficulty in breathing, without any membranous formation of the nares, without any evidence even that it had formed in the esophagus, this young woman was about to die. At two o'clock in the afternoon of the Saturday her mind was perfectly clear, her strength such that she had to be admonished not to use it. When it was proposed to do anything, to look at her throat, for example, she would jump to sit up in bed. This, of course, we forbade. *There was a blueness over the whole surface of the body*, and yet the pulse was not very feeble. Her pulse did not give warning of what was to come in five hours, and yet in that time she was dead. She did not die of dyspnœa. She did not die of the direct effects of inflammation in her throat, but of diptheritic poison, operating in some way or another apparently to prevent the free aeration of the blood, and how that could be I do not know—perhaps by some paralyzing influence on the pneumogastric nerve.

"A beautiful girl, four or five years of age, had an exudation on her tonsils which was at first treated by repeated application of a strong solution of nitrate of silver; afterward by milder local applications, as chlorate of potassa. She had but little fever, and maintained, for the most part, a fair appetite. She was most of the time cheerful and playful, though almost wholly kept in bed as a measure of prudence. The membrane forming in successive layers on the tonsils, lasted twenty days, as I have said, without extending to the air-passages or the nostrils. From the sixteenth day, she lost her relish for food. On the eighteenth, the pulse began gradually to increase in frequency without heat of skin, and without any discoverable cause advancing from eighty-five in the minute to ninety-five, one hundred, one hundred and ten. The next day it increased still in frequency to one hundred and twenty, to one hundred and thirty, and one hundred and forty; and on the third day of this acceleration, she died as the fire dies out for want of fuel. There was not the slightest dyspnœa from first to last—no hoarse cough. There was no visible hemorrhage."

C. C. Tower, M.D., of South Weymouth, Mass., in the Boston *Medical and Surgical Journal*, March 7, 1861, thus describes the disease as it prevailed in his immediate vicinity, amounting in all to seventy cases and sixteen deaths: "The patient generally feels somewhat unwell for a day or two before the affection of the throat is manifest. His appetite fails. Perhaps nausea and vomiting are the first symptoms. Adults complain of chilliness and aches in their limbs. If a child, he loses his inclination to play, and is inclined to be drowsy. There may be restlessness at night, gritting of the teeth, and feverishness. Not unfrequently none of these precursory signs appear, and if any of them occur, they are not thought of at the time, but are recalled to mind by the patient, or the parents of the child, after the more patent symptoms set in. On the second or third day, if not before, there is observed some difficulty in deglutition, and externally may be felt slight enlargement of one or both submaxillary glands, which are tender on pressure. Perhaps the first thing noticed by the parents of the child is the swelling of the areolar tissue of the throat. At this period, examination of the fauces generally reveals swelling of one or both of the tonsils and soft palate, accompanied with unusual redness of the mucous membrane. Small patches of membranous lymph, of a dirty-whitish color, are also visible. I have detected this deposit when it was no larger in extent than a split pea, but usually it is as large as a three-cent piece. I have been first called to attend a patient when the whole fauces and soft palate were covered with this exudation. Parts of the pharynx not covered with this false membrane are usually edematous and fiery red, resembling erysipelas.

"From this period the symptoms rapidly increase in severity, if not arrested. At the expiration of a week the prognosis, whether favorable or otherwise, can be determined. Death usually occurs between the end of the first and second week.

"The swelling of the throat in severe cases is very

great, so as to interfere with the venous circulation, thus producing a bloated and dusky aspect of the countenance. Breathing becomes laborious, causing the head to be thrown backward. The skin is moist, often bathed in perspiration. The pulse is rapid, soft, and small. Speech becomes lost, or audible only in whisper. The strength rapidly fails. Expectoration, at the end of a week, is quite profuse. Large flakes of fibrine, perfect castings of the air-passages, may be expelled by coughing.

"There is an odor, characteristic of this throat affection, sometimes so intense as to pervade the whole apartment.

"Death usually occurs from exhaustion of the vital forces. Frequently the little patient lies several hours in a half comatose state before life ceases. There is much suffering from dyspnœa in severe cases, when symptoms of croup manifest themselves.

"Starting at the pharynx, this disease extends upward into the nasal openings of the frontal sinuses, backward into the eustachian tubes, and downward into the trachea and bronchi, and, as I have reason to believe, into the alimentary tract."

Professor George B. Wood, M.D., of Philadelphia (Wood's Practice of Medicine), who treats of diptheria under the name of *pseudo-membranous inflammation of the fauces*, has very well collated and arranged the symptoms, as they are manifested in the great majority of cases: "The disease commences with some redness of the fauces and uneasiness, such as occur in ordinary sore throat, but usually in a less degree. This condition lasts but a very short time before the exudation commences; and, when first seen by the physician, the surface almost always exhibits small, irregularly circumscribed, whitish, yellowish-white, or ash-colored patches, sometimes seated in a portion only in the fauces, sometimes scattered here and there over almost their whole extent. These patches bear no inconsiderable resemblance to superficial sloughs, or to the surface of ulcers, for both of which they have not unfrequently been mistaken; but it has been shown, by the most careful microscopic observations, that they consist of a concrete exudation similar to false membrane, and that the surface of the membrane beneath them has not necessarily undergone any loss of substance, unless of the epithelium. Sometimes, however, ulceration is found to have taken place beneath them. In some instances the patches are translucent. Their consistence is various, occasionally pultaceous, but more frequently somewhat dense and even tough. The membrane around them is inflamed and reddened, and the tonsils are usually more or less swollen, as are frequently also the cervical and submaxillary glands, and sometimes even the parotids. Examined by the microscope, they have been found to consist mainly of interlacing fibrils, with molecular granules, epithelial cells in different stages, and often pus or blood corpuscles.

"In mild cases, such as often occur sporadically, the patches are few, more regularly circumscribed than in the severer forms, and not disposed to spread; while there is little tumefaction either of the tonsils or the external parts, and little or no fever. They are apt, however, to be attended with much pain in swallowing. In the severer cases, the patches spread with greater or less rapidity, sometimes in the course of a few hours coalescing and covering the whole fauces, but more frequently advancing rather slowly, and leaving portions of the membrane uncovered. There is commonly more of the exudation on one side than on the other; and on that where it is more abundant the swelling of the tonsils and external parts is greatest. The deglutition now becomes more difficult, and liquids often return by the nostrils in attempts to swallow. The patches, soon after they are completely formed, begin to be removed, sometimes separating in strips, sometimes softening and mixing with the fluids of the mouth, and in a few cases disappearing by absorption. They are often renewed, occasionally several times, each time becoming whiter and thinner, till at length they leave the surface covered with a puriform mucus. The process of separation usually lasts eight or ten days. (*Guemset.*) During its progress, it is attended with the discharge of some blood and copious vitiated secretions, which occasion much hawking and spitting, and hence a very offensive odor. There is often also a flow of extremely fetid sanies from the nostrils, indicating the extension of the disease to the nasal passages. The odor of the discharges in these cases has tended to confirm the erroneous idea that the disease is essentially gangrenous. In the course of the complaint, the disposition to exudation often travels downward, and the larynx, trachea, and even bronchia become lined with false membrane, which obstructs respiration, and often leads to fatal results. This extension of the disease constitutes, indeed, its chief danger. It may come on at any period, from the first appearance of the patches to the seventh or eighth day, and is indicated by those changes in the voice and respiration which characterize pseudo-membranous croup. A distinguishing characteristic of this diptheritic affection, showing that it is connected with the state of the system at large, or of the blood, is that it is disposed to appear on most other surfaces which may be excoriated or suppurating. The mucous membranes to which the air has access, and the skin, are peculiarly liable to be affected; but the mucous lining of the stomach and bowels is remarkably exempt."

PATHOLOGY OF DIPTHERIA.

I employ the phrase, pathology of diptheria, in deference to "established usage," rather than in obedience to scientific propriety. The "pathology of disease," though an expression very frequently occurring in medical literature, is as nonsensical, considered in the light of true science, as is another technicality quite as commonly found in the medical phraseology of the day, to wit, the "physiological effects of medicines." When it is considered that medicines are confessedly, in their relations to the vital organism, absolute poisons, the absurdity of the word "physiological" is sufficiently manifest. And when it is understood that disease is pathology and pathology is disease, the pathology *of* disease must be regarded as something akin to a "rhetorical flourish" or a "glittering generality."

But medical technology must of necessity be in harmony with the doctrines on which it is predicated; and if these are false, the nomenclature of the so-called science can be nothing more nor less than technical gibberish—the "incoherent expressions of incoherent ideas."

And now it so happens that the medical profession does entertain and teach—as I have shown in other works—a false doctrine of the nature of disease; a false doctrine of the action of medicines; a false theory of vitality; a false doctrine of the law of cure; a false doctrine of the relations of remedies to diseases; a false doctrine of the relations of remedies to the living organism; a false doctrine of the relations of disease to the living system; and a false theory of the *vis medicatrix naturæ;* and these facts necessitate a false technology.

The term pathology, when applied to disease, should be superseded by the word *nosology*, as this means the classification and arrangement of disease—the relation, so far as morbid conditions and actions are concerned, of one disease to another. If one should employ the term *diseaseology*, or the phrase, the *disease of disease*, or the *pathology of pathology*, he would be accused of misusing language very nonsensically; yet these phrases are not a whit more absurd—not to say ridiculous—than are one half of the technicalities to be found in medical books.

A good illustration of this subject is found in the writings on materia medica and therapeutics. For example, Pereira, in his elaborate work (Materia Medica and Therapeutics), under the head of the Physiological Effects of Corrosive Sublimate, says: "When growing plants are immersed in a solution of this salt, a part of the *poison* is absorbed, a change of color takes place in the leaves and stem, and *death is produced.*" Is not death a queer "physiological" result? I venture the opinion that no man alive, whose reasoning powers in relation to medical subjects had never been twisted, distorted, perverted, subverted, introverted, and retroverted by a medical education, would ever suspect this result to be physiological! He would certainly judge it to be just the contrary—morbid, *pathological.*

Pereira says further: "On dogs, cats, horses, rabbits, and frogs, experiments have been tried with bi-chloride of mercury, and it has been found to exercise a *poisonous* operation." And so its "physiological effects" are incomprehensible on animals as on plants. It simply poisons them to sickness or death, and this is exactly the opposite of any thing or process to which the word physiology can be properly applied.

But how is it with man? Pereira says: "Corrosive sublimate causes, when swallowed, corrosion of the stomach; and in whatever way it obtains entrance into the body, irritation of that organ and of the rectum, inflammation of the lungs, depressed action, and perhaps also inflammation of the heart, oppression of the functions of the brain, and inflammation of the salivary glands."

If this be physiological, physiology is a different thing from what the dictionaries define it to be. It is there called the "science of life." But medical authors would have it the *process of death*. It is there explained to be the doctrine of the *normal actions*. But our medical books make it the doctrine of abnormal conditions.

Physiology comprehends simply and solely the vital functions in their normal exercise, as manifested in the nutrition, development, and growth of the body; but medical men misapply the term to its morbid processes, and so confound all distinctions between food and poisons, between health and disease, between normal function and remedial effect, between the *vis conservatrix naturæ* and the *vis medicatrix naturæ*—in a word, between physiology and pathology.

Dunglison tells us in his Medical Dictionary that pathology has been defined the *physiology of disease*.

Now, disease has no physiology. It is the very opposite of physiology. It is pathological from first to last. It is *disease!* The term, "disordered physiology," as employed by Dr. Good (Study of Medicine), is appropriate, for it means, simply, *abnormal* or *morbid action*, and this is pathology, disease. It is quite as absurd to apply the term physiological to diseases and poisons as it would be to apply the term pathological to food and health.

What is diptheria? Medical writers agree that it is an inflammation, or a fever, or both! But what is inflammation or fever? This problem the profession has not yet solved. It professes to understand only the forms and features, the phenomena of inflammation and fever; but of their real character or essential nature it teaches, and books confess that they know, nothing. And why should they, so long as they can not explain what disease itself is?

Says Professor Gross: "Of the essence of disease we know very little; indeed, nothing at all." The conclusion follows by irresistible logic that, if the medical profession can not understand the nature of disease as such—if it knows not what disease is, it can not in the very nature of things understand the nature of any particular form of disease; *ergo*, the profession knows nothing of the essential nature of diptheria, and this may account for its unsuccessful treatment of the malady.

The relation of diptheria to croup and to malignant scarlet fever has been much discussed through the medical journals, some authors regarding diptheria as identical with that form of scarlatina in which the excretion of morbid matter is mainly determined to the throat, with little or no cutaneous eruption, while others regard it as differing from croup only in the fact that the inflammatory action affects principally the mucous membrane of the throat, instead of that of the trachea or windpipe, as in the case of croup.

Whatever may be true in theory, the facts are sufficiently obvious. It is true that in the cases of membranific inflammation, to which the term diptheria is usually applied, there is an exudation of fibrinous material—coagulable lymph—on the mucous surface of the throat, quite analogous and frequently identical in character with that which takes place on the mucous surface of the trachea in croup; and in some cases extending, as we have seen, into the windpipe and esophagus. It is also true that *scarlatina maligna* is characterized by ulcerative inflammation of the throat, instead of the exudation which produces the false membrane. And it is equally true, moreover, that in some cases of diptheria ulceration does take place beneath the membranous formation, while in some cases of putrid sore throat there are patches, more or less extensive, of diptheritic or croupal excretion. For all practical purposes, therefore, we may regard diptheria as combining the morbid conditions of both croup and malignant scarlet fever, one or the other being more prominent according to the condition and habits of the patient in whom the disease occurs.

There is much discrepancy among the authors as to the symptoms of diptheria which are supposed to identify it with croup on the one hand, or malignant scarlatina on the other. Some authors have noticed a scarlet eruption of the skin accompanying the throat-affection, and the putrescent condition of the whole system, attended with fetid breath and foul excretions, as in malignant scarlet fever; while other authors describe it as being entirely free of offensiveness and eruption.

Both sets of authors are correct in their facts, yet mistaken in their theories. The disease occurs in persons of very different dietetic and other personal habits, and in very different conditions of system, so far as grossness of blood and exhaustion of vital power are concerned. These facts, which are of the utmost importance in enabling us to understand the rationale of the various forms of diptheria, seem to be wholly overlooked and unthought of by the authors who have written on the subject.

Every febrile disease of the continued type—and diptheria is a continued febrile disease; that is to say, a local inflammation *essentially* accompanied with a constitutional fever—is either of the *inflammatory* or *typhoid* diathesis; and if of the typhoid diathesis it is either of the *putrid* or *nervous* form. Diptheria is always of the atonic, low, or typhoid diathesis; and never attended with high or entonic fever. And if the condition of the patient be very gross, the disease will present the putrid phase of fever, usually called *typhus*, or *typhus gravior* in medical books; but if the patient be in a less gross and more debilitated state, the disease will be of the nervous form, the *typhus mitior* of the older authors, the *typhoid fever* of modern authors, and the *enteric fever* of Wood and others.

But our medical authors are wholly at fault as to the causes of these distinctions. They are usually attributed to some specific property in the morbific material which induces the malady, or to some quality inherent in the disease itself, as though this were a thing or entity outside of the living organism and capable of assuming a variety of shapes, whereas the real cause is to be found solely in the condition of the patient. In some places, where nearly every case of diptheria has terminated fatally, the disease is said to have been of a very malignant type; and in other places, where few deaths have occurred, it is said to have been of a very mild type; as though the disease had an existence and a character before it "attacked" the living organism.

Such is the medical science of the nineteenth century, but it is a delusion. The disease, so far from existing as an entity outside and independent of the living system, so far from being a thing acting on or attacking the system from without, is really, as is every other form of disease, the action of the vital organism itself. It is the living system in the act of expelling impurities—a process of purification. And malignancy is determined, both in nature and degree, by the condition of the system, and always exists in the exact ratio to the grossness or putrescency of the blood, or the debility or exhaustion of the nerves. When diptheria, therefore, occurs in persons of so gross and putrescent a condition of blood as to induce the putrid form of fever, the breath of the patient may be very fetid, and the discharges very offensive, and the odor of the patient's room very disagreeable; while, if the disease occurs in persons of a feeble but not gross condition, although it may be equally malignant and equally fatal, it will have none of the peculiar evidences of putrescency of the fluids which characterize the other form.

Diptheria, putrid sore throat, and croup, though usually distinct in diagnosis, may run into each other, as it were, by such imperceptible gradations that it is sometimes difficult to draw the line of demarkation and tell where one ends and the other begins, diptheria holding the intermediate relation, and combining in itself more or less, in different cases, the conditions of exudation and ulceration; while croup and putrid sore throat represent, more distinctly, the membranous and the ulcerous forms of inflammation, as it is presented in a low atonic or putrescent state of the system.

Some authors have seemed to confound diptheria with quinsy—*tonsillitis;* but in the latter case the swelling and redness of the tonsils, with difficult deglutition at the outset, and the absence of all evidences of fibrinous exudation, are sufficient to enable the careful observer to distinguish between them.

Authors have disagreed also respecting the febrile or non-febrile character of diptheria; and we are gravely assured by some authors who profess to have had much experience, that it is scarcely ever febrile; while other authors as gravely talk of the disease " becoming typhoid," or of " typhoid symptoms supervening." The truth is, the disease is *always* febrile, as are all acute visceral inflammations, although, in many cases, as in all visceral inflammation of low diathesis, the hot stage of the febrile paroxysm may be very slight and scarcely observed at all, or entirely overlooked. And the fever is *always typhoid* from first to last, so that the phrases " running into typhoid," " typhoid supervening," etc., indicate an erroneous view of the character of the malady.

The following article appeared not long since in the New York *Commercial Advertiser*, and it represents very well many errors, both in pathology and therapeutics, which are entertained by the great majority of the medical profession. I copy it for the opportunity it affords for corrective criticism.

"This disease, though in many respects resembling croup—and in certain others, quinsy—is distinguishable from both by certain well-marked characteristics. Like croup, it is accompanied by the formation of a false membrane in the windpipe, which, if left to itself, accumulates till the air-passage is closed and death ensues. But the false membrane of croup is an exudation of natural lymph from the vessels of the mucous membrane stimulated to excess by high febrile condition of the tissue; while, on the other hand, diptheria is scarcely ever febrile in its pathology—and its pseudo-membrane is the result of a *sloughing off* rather than an *exudation* of the mucous coating. Croup belongs to the inflammatory type of diseases— diptheria, save in exceptional cases, does not. In croup, the breath of the patient is usually untainted. In diptheria, the breath is characterized by a peculiar and sometimes almost intolerable fetor. The lymphatic discharges of croup are seldom acrid. The discharges from the nose and mouth of a diptheritic patient are ichorous and excoriating to the highest degree. Croup is not particularly prostrating to the general strength of the person attacked by it. Diptheria is invariably accompanied by extreme debility, and a loss of muscular as well as nervous tone, which often continues for months after the immediately dangerous symptoms have been overcome. Finally, diptheria is contagious —croup is not.

"It will be seen from these details that diptheria and quinsy have more intimate points of resemblance than diptheria and croup. In certain cases this resemblance is greatly increased by a complication of the pseudo-membranous symptom of diptheria with malignant inflammation of the tonsils. Still the pseudo-membranous symptom is of course always sufficient to distinguish it from quinsy.

"It is not probable that diptheria is a new disease. The scientific accuracy of medical terms has made such rapid progress during the last half century, that the physicians frequently find the data of diseases, as reported thirty years ago, unavailable through vagueness for the purposes of an indicative experience. Nevertheless, from all that can be ascertained on the subject, the identity of diptheria with the "putrid sore throat," which made such fearful ravages in Albany and other places a quarter of a century ago, seems very probable. That malady was characterized by marked typhoid symptoms, and this indication has its counterpart in the extreme prostration of diptheria. If we recollect rightly, not a single case of the old putrid sore throat, which received the then universal depleting treatment of calomel and blood-letting, ever recovered from the disease. At the present day, nobody in his senses would think of letting blood or giving exhaustive medicine for diptheria.

"The treatment of the disease proposes to itself two ends:

"1st. To evoke and sustain all the natural vital forces of the patient.

"2d. To rid the air-passages of the false membrane.

"For the attainment of the first end, nutritious, digestible *food*, being the most natural, is, of course, also the best means. Strong beef tea combines all the most desirable elements for such a purpose. It should be given from the earliest stages of the disease; and when, as sometimes happens, the fauces become closed by the disease, or the parts become too painful to admit of swallowing, it is still to be given in the form of anal injections. Brandy, in judicious hands, is another stimulus of the highest value in diptheria. Iron in various forms has been administered with great success. Perhaps its most efficient form, as determined by late experience, is the sesquioxide. Its effect seems to be two-fold—sustaining the general strength of the patient, and assisting the tendency of the mucous membrane to throw off and eject the diptheritic slough.

"For the attainment of the second end innumerable methods have been proposed, and some of them have been very successful. Occasionally the use of the sesquioxide above mentioned has been successful, in co-operation with the forces of Nature, to produce the rejection of the membrane as fast as it accumulated and before it was indurated sufficiently to exclude all air from the lungs Sometimes it has been found possible to detach and pull out the membrane by means of a hooked or forcep-shaped instrument, though this operation has been known to result in dangerous hemorrhage. But by far the most successful treatment for relieving the diptheritic patient of the false membrane is that recently discovered by Dr. Lewis A. Sayre, of this city. His method is one of those admirable attainments of the highest class of inventive genius which, from their extreme simplicity and obviousness, awaken in every mind the remark—'Why, I might have thought of that myself!' Yet nobody ever does think of it till the inventive genius happens to show him the way.

"The method of Dr. Sayre was the result of the following observation. He noticed that if the discharge of diptheria was expectorated upon any dry and warm substance—such as the side of a stove, for instance—it immediately became a tough pellicle, like a shred of gold-beaters' skin. But if the expectoration fell into a vessel containing warm water, it remained liquid and limpid, like ordinary thin lymph or mucus.

"It now occurred to Dr. Sayre, that if from the first stages of the formation of the false membrane a hot and humid atmosphere could be kept in contact with it, it would remain as soluble as in this last-mentioned case, and be easily ejected through the nose and mouth like common mucus.

"Several means of procuring this contact suggest themselves. The well-known plan of inhalation from the spout of a tea-kettle, and the ordinary vapor-bath, are among these. But the former is evidently impracticable with those young children who are the most frequent sufferers from diptheria. They can not be made to keep their mouths in position over the narrow steam surface of a kettle. The vapor-bath is relaxing to the general system, and can not be thought of in a case which, like diptheria, requires every tonic and stimulant that can be made available. Moreover, it would be manifestly impossible to continue the patient in a vapor-bath through a period as long as the membrane is accumulating.

"Dr. Sayre finally adopted this method. Having put the patient in a tightly shut room, he had a flat-iron heated to as near the white heat as possible. He suspended it over a pail in the sick-room, and kept the attendants pouring water on it till it ceased to evaporate every drop that came in contact with it. As soon as the iron was cooled down to such a degree that any surplus of water remained unevaporized, he replaced it with one freshly heated. He thus kept the room as full of steam as was consistent with comfortable breathing—at a temperature of 80° F. This process was continued for several hours; during which not only the freshly sloughed membrane was constantly being expelled in liquid form through the nose and mouth, but membrane previously indurated in the trachea became soluble and was ejected in like manner. Meanwhile he kept up the strength of the patient by the above referred to means of beef and brandy.

"The result of this treatment was an entire expulsion of the slough, and, eventually, the complete cure of a case which had previously been abandoned as too desperate for even the dernier operation of tracheotomy.

"Undoubtedly the means of evaporation for this purpose will hereafter be simplified by the discovery of the method. He has plans at present under consideration by which the process may go on independently of the laborious and sometimes unreliable co-operation of attendants. Still, it is now a fixed fact that we have made the great and conclusive step toward a certain cure of diptheria. Also, that we owe that fixed fact to Dr. Sayre.

"The utmost care of the patient for weeks after the immediately dangerous symptoms have disappeared, is necessary to prevent a subsidence into the diptheritic state. Even where there is no return of the sloughing tendency, the general prostration of the system is usually so extreme, that the most nourishing tonics and stimulant treatment are called for to ward off a naturally supervening attack of typhoid or low nervous fever, rapid decline, or chronic debility. There is perhaps no form of disease known to the children's practitioner in which skillful hygiene and home-treatment are more imperatively demanded to follow up and perpetuate the results of medical effort. If possible, the greatest care must also be taken during the period of fetid discharges to separate the remaining children of a family from the diseased one, for, as we have above observed, this stage of the disease is quite infectious.

"We notice that diptheria is again beginning to manifest itself as an epidemic in some of the rural districts of New York and the neighboring States. The words we have said may be of still more use in a city like ours, where life is so closely packed, infection and death so easy. But of these latter evils there is no need. Cure is now measurably simplified—prevention simpler still."*

"Diptheria is *scarcely ever febrile* in its pathology." On the contrary, diptheria is *always febrile* in its pathology. "Croup belongs to the inflammatory type of diseases; diptheria, save in exceptional cases, does not." Both croup and diptheria are *always inflammatory* and *always febrile*, each disease consisting essentially in a local inflammation *and* a constitutional fever. So far as inflammatory *type* is concerned, the word is misapplied. Type, when properly employed, pertains to the periodicity of the febrile paroxysms, as the continued remittent and intermittent *types* of fever, and not to the entonic or atonic diathesis which characterizes them, nor to the inflammatory or non-inflammatory nature of a disease. The terms *type* and *diathesis* are employed quite promiscuously by modern medical writers, as are, indeed, a hundred other tech-

---

* In the above article are many errors, both theoretical and practical.

nical words and phrases, but it is for want of clear and correct ideas.

The breath and discharges are generally more or less fetid and acrid in diptheria, as already remarked, according to the greater or less grossness or putrescency of the patient; but in some cases which I have seen, these symptoms were entirely wanting, while there are cases of true croup in which the excretions are foul and offensive.

"Croup is not particularly prostrating to the general strength of the person attacked by it." Such language indicates the false notion which the medical profession entertains of the nature of disease—an error which I have been combating in books and in lectures for a dozen years, and which I have made a prominent topic in all of my works and writings. But as the limits of this work will not permit me to discuss the subject at length, I can only refer the reader to my large book, the "Hydropathic Encyclopedia," and to some of my smaller works, particularly "Water-Cure for the Million," "Principles of Hygeio-Therapy," and "The Alcoholic Controversy," for a full exposition of the theory involved.

The idea that diptheria "prostrates the person attacked," implies that the disease is a separate and distinct entity from, and an existence outside of the living organism; and this absurd theory is the basis of all the false medical science and bad medical practice in the world. The truth is, disease—all disease—is simply the action of the living system in self-defense—a process of purification—*a remedial effort.* When this action occurs, when this struggle begins, the system may be in a condition of great obstruction or of extreme exhaustion, corresponding with and occasioned by the personal habits, manner of life, occupation, exposures, etc., of the patient; and the form (not type) of the disease will be putrid or nervous, as one or the other of these conditions is most prominent, and its form or tendency (not type) will be mild or malignant, not according to some imaginary specific character of the morbid entity, or what authors so vaguely denominate "epidemic constitution," but according to the greater or less putrescency or debility of the system. When medical men recognize these distinctions, they will have a much more rational pathology, and a vastly more successful practice.

"The loss of muscular as well as nervous tone," and the numerous *sequelæ*, in the shape of chronic diseases, which so generally follow the "attack" of diptheria, are, in my opinion, chiefly the effects of the drug-remedies—in other words, drug-diseases.

So far as the treatment recommended in the preceding article is concerned, I will merely remark in this place that there are some things in it to commend and some to condemn, reserving the further discussion of the matter until I come to consider the therapeutic application of the principles I shall endeavor to establish.

Dr. Wood remarks: "In good constitutions the fever is usually sthenic; but sometimes, especially when the disease prevails epidemically, it has a typhoid or malignant character, and this condition of the system *reacts on the local affection.*" The expression which I have italicized is utter nonsense. The constitutional affection or condition does not act nor react on the local affection, nor does the local affection act nor react on the constitutional condition. The local affection is an inflammation, and the constitutional affection is a fever. Both together constitute the disease. The fever is *never* sthenic; but in good constitutions, that is to say, in constitutions not very gross nor very much enfeebled, for the reasons already assigned, the fever will not be *very low*, but still it will be typhoid. And the inflammation, in diathesis, always corresponds with the fever; hence it is always low, passive, atonic, typhoid, asthenic.

The doctrine is everywhere recognized in medical books, that a local inflammation and the accompanying fever may be of opposite diatheses, so that the remedies which are demanded by the local condition are injurious to the general system, and *vice versa*. This is one of the most pernicious of the many fallacies of a false medical system, as it inevitably involves the practitioner in the inexplicable muddle of "indications and contra-indications," and necessitates the administration of remedies of the most conflicting "*modus operandi*," and insures the death of a large proportion of the patients.

The inflammation and the fever—the local and the constitutional affection—in diptheria, as in all diseases, always correspond in character, in diathesis; and the local affection never requires that treatment which aggravates the constitutional condition, nor does the general system ever demand any remedy or plan of treatment which is not also best for the local affection; and when it is understood that the inflammation of the throat and the fever of the system are parts of one and the same disease, the idea of one "reacting" on the other is sufficiently absurd.

Dr. Wood remarks further: "In the malignant cases the system is probably under some poisonous influence, superadded to that of the local affection."

Our author does not seem to have the remotest idea of any rationale of malignancy. Malignancy does not imply any "superadded" poison, but a great amount or quantity of poison, or a feeble organism.

The distinction between the diptheritic exudation and apthous sore mouth, or *thrush*, is well explained by Dr. Wood: "In the thrush, the white coating appears first in separate points, which afterward coalesce; is formed upon the surface of the epidermis, or at least not beneath it; may be readily removed without affecting the integrity of the mucous membrane, or causing the least hemorrhage, and, when examined under the microscope, is found to contain abundantly a peculiar fungous plant. The diptheritic exudation forms in patches, *beneath the epidermis;* adheres strongly to the membrane, so that it can rarely be detached without causing the extravasation of some blood; and under the microscope exhibits the ordinary constituents of false membrane; namely, interlacing fibrils, molecules or granules, and exudation or pus corpuscles. The exudation in scarlatina occurs generally first in points, like the thrush, is much less cohesive than the diptheritic, less adherent to the mucous membrane, much less disposed to spread into the larynx, and also less disposed to make its appearance upon surfaces elsewhere that may be excoriated."

Dr. McDonald, of Bristol, Eng. (*Braithwaite's Retrospect*, Jan., 1860), says, in relation to the identity of diptheria and malignant scarlet fever: "There has been considerable confusion with respect to scarlet fever and diptheria. Some have contended for the identity of the two, maintaining that those cases in which no rash appeared were to be considered as 'suppressed scarlet fever.' To combat this view, it will be sufficient, I think, to draw attention to the great difference in the symptoms I have described from those of scarlatina, and to state the fact of its having been my painful experience to have attended families, some members of which have been swept off by scarlet fever *with diptheria*, while other members, who had previously suffered from scarlet fever in a severe form, were now attacked with true diptheria. That scarlatina invites diptheria is very manifest, but that the diseases are perfectly distinct and different is equally certain."

In another article, in the same number of *Braithwaite*, J. C. S. Jennings, Esq., of Malmesbury, Eng., says of the diptheria as it appeared under his observation: "At the first outbreak of the disease no cases of scarlatina had appeared in the neighborhood, nor were there any until the second outbreak during the month of January in this year, when a few cases of diptheria occurred; but scarlatina maligna ran through several families. In those cases, however, in which the rash was well developed and not suppressed, there was little or no throat affection; and *vice versa;* and when the tonsils *were* affected, there was not the peculiar leathery exudation of diptheria."

Thomas Neckstall Smith, Esq., before quoted, in reference to the "type" (diathesis) of diptheria, remarks (*Braithwaite*, part 46): "Have we seen this disease before? and what is its nature? In answer to the first question, I can say confidently, that during a period of upward of thirty years' practice I had seen no case of diptheria until 1857. I had read Bretonneau's earlier papers many years since, and should have recognized the disease had it presented itself. Of its nature it is less easy to speak. It is evidently, I think, a blood-disease, and not merely a local one. But what is the nature of that abnormal condition has yet to be explained, or, rather, I fear, has yet to be discovered.

"In observing the progress of this epidemic I have been instinctively led to reflect on the altered type of disease in general. I have, myself, no doubt of that alteration in the type of disease, observed since the year 1832 in England. From that date there has been a departure from the old sthenic type, and this has been more pronounced the last few years, until at length a genuine sthenic form of illness is almost, if not quite, unknown among us. We have, instead, low types of inflammation, low forms of cutaneous diseases, low types of fever, having more and more a tendency to the remittent form; and a very marked increase in localities where it was before almost unknown, and where no known causes have arisen to occasion it, of intermittent fever. What was before a mere chill, a slight cold, thrown off with the first reaction, becomes now an attack of ague."

The explanation of this change of *diathesis*, which the author before us denominates "type," is not difficult to understand, in the light of the premises I have advanced. The diathesis of disease always tends from high to low—runs down, so to speak—as the constitutional vigor of the people declines. The change is not in the disease, *per se*, but in the habits of the people. Our fathers and grandfathers, our mothers and grandmothers, when they had inflammatory and febrile diseases, manifested the high, active, entonic diathesis much more frequently than do their more effeminate sons and daughters. The lower the vital stamina, the lower will be the diathesis, because the more feeble

3*

the vital struggle, in those remedial efforts which constitute the various forms of fevers and inflammations. Is it not strange that medical men have so long looked in the wrong direction for the solution of this problem?

Mr. Smith continues: "We have abundant evidence of this depression of vital power in the general symptoms of diptheria. We have also a low type of local inflammation in unison with the general type; but why it should just now seize the throat as its local seat instead of showing itself as boils, carbuncles, whitlows, thecal abscess, necrosed bone, and in kindred forms, I do not know."

Nor will our author ever know if he forever pursues the phantom-entity which medical books denominate disease. The inflammation *chooses* to attack the throat; it *prefers* that as its seat; it *seizes* on that locality in preference to another; it *elects* to manifest itself in the form of diptheritic exudation instead of carbuncular ulceration! Such are the vagaries of learned medical men! Such is the ridiculous nonsense which makes up the chief burden of medical books, and which is called science! I am of opinion that the whole mystery lies in a false notion of the nature of disease, and that the reason why disease assumes one form instead of another is, because the living system, under all the circumstances, can best depurate itself of impurities by the actions which constitute the leading symptoms of the existing disease; or, at least, can not, under the circumstances, make any other or different effort. To illustrate: if the system has sufficient power to determine the remedial effort chiefly to the surface and maintain it there, the fever, or the inflammation, or both when they co-exist—the diathesis—will be high, entonic, inflammatory, dynamic, or asthenic; but if this is not the case, if the system is too gross or too feeble, the remedial effort will be directed mainly from the surface, and the diathesis will be low, atonic, typhoid, or asthenic.

That diptheria and other forms of throat inflammation run into each other, so to speak, by imperceptible gradations, is apparent to all who have had much experience in treating these maladies.

Says Dr. Edward Ballard, in an account of diptheria and epidemic sore throat, as they prevailed in the parish of Islington in 1858–'9: "The prevalence of sore throat not diptheritic in character, during the past year, has been matter of general remark. Many, if not most of these throats, exhibited some approach to the color of the mucous membrane when about to become the seat of diptheritic exudation. These sore throats appear to bear about the same relation to diptheria as diarrhea bears to cholera in epidemic seasons. Just as in any cases of diarrhea, in an epidemic period, it is impossible to predicate that it will not pass into cholera, if neglected, so, in the ordinary sore throats which have lately presented themselves, no one would be bold enough to assert that any one might not before long exhibit the characteristic symptoms of true diptheria."

One more theory remains to be considered before dismissing this branch of our subject. Dr. T. Laycock, of Edinburgh, Scotland, has put forth the theory (*Braithwaite*, July, 1859), that the diptheritic exudation depended on a *parasitic fungus* in the *oidium albicans*. This opinion is undoubtedly erroneous, as Dr. W. R. Rogers has explained: "The oidium albicans is not found in diptheritic exudation, unless in exceptional cases, and then only because the membrane has taken on an acid, putrefying change, this parasite requiring an acid, decomposing pabulum whereon to flourish, as is well proved by Berg and Gubler. In France, all know that this fungus distinguishes the *pseudo* from the true diptherite, the microscope being its test. Wherever this oidium is found it is muguet or thrush, *plus* whatever disease, acute or chronic, it may be, as I stated in my paper read before the Medical Society. I have only found the oidium in one out of fourteen specimens, and this was fifty-six hours after the patient's death, though carefully examined twenty-four hours before. I may add that the *leptothrix buccalis* mentioned by Dr. Wade is constantly to be found in the buccal mucus of healthy persons, if properly searched for. Dr. Harley, of University College, has stated to me that fatty acids are frequently mistaken for this fungus. True diptherite, in all the specimens I have examined, is a granular and cellular exudation, with some epithelial mucous corpuscles, and sometimes there may be found with it pus and blood-cells. I have but rarely distinguished fibrillæ, or what looked like these. Under the exudation, the sub-mucous tissues and mucous membranes are usually thickened, and the mucous follicles are enlarged and filled with the same matter, which can be squeezed out, and from which the exudation seems to be produced; but the cause of this change of mucus into membrane I do not desire at present to enter upon."

As Dr. Laycock has published a reply to Dr. Rogers, through the London *Lancet* for January, 1859, my work would be incomplete without it, although I do not regard it as sustaining his position, while it may not be very interesting to the reader:

"If I understand Dr. Rogers' views aright, as reported in the *Lancet* of the 22d inst., he not only thinks diptheria to be a blood-disease (which seems a probable theory), but that, *as such*, it can not be a parasitic disease. Comparative pathology teaches, however, that this conclusion is, at least, doubtful. The muscardine (an epizootic disease of the silkworm) is due to a species of fungus like that which infests the potato, called, after its discoverer, the *Botrytis Bassiana*, and the sporules are described as being *reproduced in the blood* of the insect *when it becomes acid*, while the filaments and mycelium appear on the respiratory surfaces—that is, at the outlets of the tracheal tubes. (Compare the engraving of the blood-appearances in M. Ch. Robin's valuable 'Histoire Naturelle des Végétaux Parasites,' etc.) Again, the fungus of the common house-fly, named mycophyton Cohnii by Lebert, after Dr. Cohn, its first investigator, is a mold or oidium found in the blood, abdomen, and sometimes in the intestines of the insect at the beginning of autumn. (Lebert, Virchow's 'Archiv.,' etc., vol. xii., 1857.) Its first symptom observed is a milky appearance of the blood. It is found in the blood in all stages of development, from the simple minute spore, or cell, to the full-grown mycelium. It is found in like manner in the fluids of the intestines, and appears externally as a mold. Flies thus affected may be often seen sticking, with outstretched wings, to the window-panes, at the end of summer and beginning of autumn. These are by no means solitary instances of parasitic blood-disease. Indeed, hæmotophyta, as Lebert terms these microscopic blood-parasites, infest the blood of several classes of insects. References are given by Lebert *loco citato*. The same fact also holds good as to the vegetable parasites. The common wheat bunt attacks the wheat, and makes it *look* and be sickly, when not the slightest trace of fungal thread can be found; yet it is quite certain that something capable of reproducing the species is present at the time, either in the intercellular passages or protoplasm. This I state on the authority of Mr. Berkley. ('Introduction to Cryptogamic Botany,' 1857, p. 65.) That eminent observer is also of opinion that the Botrytis infestans is the fungus which is the cause of the potato disease. He says a crop may be seen to grow in a few hours from the cut surface of a diseased potato, even although the foliage exhibited no traces of the parasite; and that the walls of the cavities of the carpels of the tomatoes are often covered with the fungus, though there is no communication with the external air. These are facts which ought to make us hesitate, at least, in coming to the conclusion, in the absence of all inquiry, that a parasitic disease can not be a blood-disease in man. The same kind of objection applies to the conclusions drawn from microscopic investigations by Dr. Rogers and Dr. Harley. A hundred examples of wheat infected with the tilletia caries (the bunt) might be examined in succession, or even a thousand, and no fungus detected; but that would obviously be no proof that the diseased condition of the grain was not due to the parasite. It would simply signify that the diseased grain had not been examined at the proper stage of the development of the fungus. And I think the fact stated by Dr. Harley and Dr. Rogers, as to one of the twelve cases they examined, that the oidium albicans was developed twenty-four hours after no trace of it could be found, is significant of what may be, and I think is, the rule in the living body—namely, that certain conditions are necessary as to development, food, temperature, and habitat, for the complete evolution of these organisms. There is no doubt that an acid condition *accompanies* the production and growth of the oidium in muguet, and of vegetable parasites on the skin in skin diseases; but it is not so clear that the acid is the *cause* thereof. On the contrary, we know that the production of acid is itself due to fungi, as in the acetous fermentation. Dr. Lowe, of King's Lynn, differs from Gubler and others as to this acid theory, and I would particularly call Dr. Harley's attention to the account of Dr. Lowe's interesting experimental researches on these parasitic fungi, published last year, in the 'Transactions of the Botanical Society of Edinburgh.' The title of Dr. Lowe's paper is significant of the caution with which microscopic researches should be made. It runs thus: 'On the Identity of Achorion Schönleinii and other Parasites with Aspergillus Glaucus.' Dr. Lowe believes he 'raised' aspergillus glaucus from the parasitic fungus (the achorion of a case of porrigo lupinosa, treated in the Royal Infirmary here), and he got good yeast (torula cerevisiæ) from both the aspergillus and penecilium, which might, therefore, be got from the favus-fungus. Dr. Lowe infers, in fact, from his experiments, that all the fungi which produce skin diseases are referable to these two genera which produce yeast; and conversely, that yeast may, under favorable circumstances, produce skin disease. The *leptothrix*, so common on the foul tongue, is probably to be classed with these favorable forms. These statements show, at least, how much is yet to be done in natural history before the true morbific action of these parasitic fungi can be determined. One thing, however, is certain, that the parasites of the potato, vine, apple, and silk-worm, all prevailing simultaneously, are almost identical with the oidium albicans; and considering how readily a slight difference in the form of these minute organisms may be induced by differences in the food or habitat, it is probable that they are really identical in origin; and this coincidence of spread can not but awaken a strong suspicion as to the relationship of the cause of diptheria to that of the epidemics of the silk-worm, vine, potato, etc.

"That these parasites are sometimes powerful irritants of the living tissues is, I think, fully established, both from the history of muguet and other circumstances. And although French writers speak of *pseudo* diptherite, the accuracy of the term may be questioned, for the exudation appears externally on ulcerated or exposed surfaces as well as internally in both muguet and diptheria alike. An interesting case of vaginal blennorrhea, due, probably, to oidium albicans introduced from without, may be found in Virchow's 'Archiv. für Physiologie,' vol. ix., p. 466. The case is communicated by Dr. E. Martin, of Jena. The labia were swollen; the vagina of bright red, studded with enlarged papillæ, and covered with star-like patches of membrane like those of the mouth in muguet, which were found to contain the oidium albicans. A patient in the next bed (both were puerperal patients in hospital) had subsequently active fever, abdominal tenderness, and oidium albicans of the mouth, with muguet. Dr. Jos. Ebert, of Wurtzburg, found the oidium albicans in the crop, stomach, and intestinal canal of a hen. The upper portion of the latter was intensely red.

"It is usual to speak of the characteristic pellicle as if it were peculiar to diptheria; but this is by no means the case. It is not unfrequently seen in cases of typhus and relaxing fever, sometimes in yellow fever, and, I believe, in all fevers. A series of carefully conducted experiments made with a thorough knowledge of cryptogamic botany, on lower animals, so as to show the real pathological origin and effects of these parasitic fungi, would be very valuable. It would be absolutely necessary, however, that the animals experimented on be first brought as nearly as possible under, and into, the same conditions as persons are in who are attacked by the disease. I am inclined to think that it would probably be shown that these parasites may act either through the blood or locally only."

THE FALSE MEMBRANE.

In all forms of disease—distinguishing *action*, which constitutes the essence of disease, from *condition*, which may be its cause, accompaniment, or effect—there is an effort on the part of the living organism to rid itself of abnormal conditions, effete matters, foreign substances, or what modern physicians have, with singular absurdity, denominated " morbid poisons;" indeed, this process of depuration, as I have already explained, constitutes the " essential nature of disease." When the process of depuration is directed mainly through the ordinary channels—the skin, liver, bowels, kidneys, and lungs—we have the simple fevers, varying in form and phenomena, in type and diathesis, according to the quantity and quality of the impurities, etc., to be expelled, and the greater or less vigor of the various organs at the time ; these circumstances affording the rationale of the distinctions of simple fevers into inflammatory, bilious, typhoid, continued, remittent, intermittent, ephemeral, etc. When the noxious materials are determined almost wholly to the surface, and are of a nature to be eliminated only, except to a small extent, through the cutaneous emunctory, we have the various forms of eruptive fevers—small-pox, measles, scarlet fever, erysipelas, miliary fever, etc. In these eruptive fevers there is, in almost all cases, more or less expulsion of morbid matter upon the surfaces of the mucous membranes, constituting an eruption, exudation, or ulceration of the part. In one form of scarlet fever—*scarlatina maligna*—the determination of morbid matter is mainly to the throat, presenting what has frequently been called "putrid sore throat." In some conditions of the system the noxious materials are thrown upon the mucous membrane of the trachea or windpipe, and so charged with the fibrinous element of the blood—coagulable lymph—that, after being removed from its normal relations, it concretes into a false membranous coating, constituting *true croup;* or, if the fibrinous element is incapable of being thus partially or imperfectly organized, and is expectorated as a dense, glairy excretum, it constitutes the *false* or *non-membranous croup*. When morbid matter thus affects the tonsils, or is specially determined to the mucous surfaces of the nose, or of the fauces, we have the *common quinsy*, or the catarrh, or the malignant quinsy—the "black tongue" of domestic animals. And when the fibrinous material is exuded over a greater or less portion of the mucous membrane of the mouth, whether or not involving the larynx, trachea, bronchia, and esophagus, it constitutes the disease which is generally recognized as diptheria. This exudation also occurs in some cases of diarrhea and dysmennorrhea, and in catarrh of the bladder; or rather, the inflammatory process which excretes the membranous matter is the cause of those particular forms of disease which have been termed *tubular diarrhea, painful menstruation,* and *catarrh of the bladder.* In these cases the membranous formation is usually broken up by the contraction of the parts, and expelled in fragments. In some cases, however, it has been cast off entire, and then not unfrequently mistaken for a sloughing and expulsion of the mucous membrane itself. Even old and experienced professors of obstetrics, and authors of standard works on diseases of women, very often mistake this morbid product for the "cast-off mucous membrane," when the uterus is the seat of the exudation.

Dr. Winne has collated from various writers a very good description of the false membrane of ordinary diptheria. "When the mouth is examined upon the first day of the pseudo-membranous deposit, the parts destined to become the seat of the disease present the appearance of pieces of flesh bleached by contact with boiling water; soon after there appears on the tonsils, the uvula, or the soft palate, small vesicular points of a lardaceous appearance, formed by the dissolving of the epithelium, which may readily be confounded with the minute yellow patches soon to appear. The membrane is almost invariably developed on one or the other of the tonsils, but not always, as the uvula is sometimes the original seat of the patches."

Ordinarily, at the moment of formation, or soon after, the false membrane appears under the form of a white or a yellowish-white spot, rarely gray, quite circumscribed, a little projecting at its center, and surrounded by a circle of lively red. Sometimes the false membrane is semi-transparent and forms a slight pellicle, which envelops the tonsils, through which the surface of this gland is partially visible; but it soon loses this transparency and becomes of a yellowish-white color, extending itself to the subjacent parts with greater or less rapidity, according to a variety of circumstances, and especially the kind of treatment which has been adopted. After the false membrane has developed itself upon the tonsils, it usually extends to the soft palate, the uvula, and finally to the pharynx, with greater or less facility, regularly involving these different parts in the order here indicated. This is not invariably the case, for sometimes it is developed simultaneously in several distinct points, which finally converge the one into the other, and finish by forming a continuous surface. While it is thus enlarging its boundary, the false membrane acquires an additional thickness by the crossing of successive layers, so that it is not composed of one single film, but of many, which present a varied appearance, dependent upon the place occupied by them, "sometimes appearing like a deep ulcer with a yellow base; at others, enveloping the uvula as a finger by a glove, and on the palate having the semblance of a deep hollow.

"The period between the formation of this membrane and its dislodgment is very variable—usually from one to six days. In the early part of the disease, after being detached, a new membrane forms in its place, and this may be habitually reproduced several times. When the membrane is cast off spontaneously about the sixth or seventh day, its place is seldom supplied by a new deposit; and about the tenth day the patient is convalescent. When the case terminates fatally, the original inflammation extends to the air-passages, and not unfrequently to the nasal cavities, which likewise become the seat of a pseudo-membrane, greatly augmenting the sufferings of the patient and the gravity of the disease, whose termination is heralded by the fetid, sanious discharge from the nostrils, and symptoms of angina, which speedily supervene.

"When the termination is hastened by the supervention of gangrene, the pseudo-membrane loses its consistency, is easily detached, changes to a grayish color frequently mixed with bloody spots, and is coated with a sanious fluid which flows from the mouth and nostrils, and emits a very fetid odor. The flow of blood in these cases is sometimes considerable, and not unfrequently covers the lips and nasal cavities, in which latter the flow is often arrested by the formation of clots.

"Whatever may be the time at which the false membrane becomes detached, it generally exhibits the subjacent tissues diminished in size, and of a redness more or less intense in color. This diminution in size is especially noticeable in the tonsils and uvula. The false membrane does not always occupy the same seat."

Dr. Pichenot, in a report to the Paris Academy of Medicine, describes the local symptoms of a very fatal liptheria which prevailed epidemically in the Commune of Creusery, in 1855 : "The tonsils press upon the folds of the soft palate, and their surface is injected with a grayish deposit. But it is upon the mucous membrane of the posterior portion of the throat that the diptheritic plagues usually present themselves, and their grave condition here almost invariably presages grave and rapid disease, and not unfrequently a fatal termination. The pain in the head and neck now becomes augmented, the respiration more difficult, the face edematous, and the maxillary glands tumefied and sensitive to pressure.

"Its march is very rapid. In the space of from three to five hours the papular eminences of the throat become covered with a flocculent, transparent vail, of white appearance. Generally, not more than one half of the guttural cavity is at first invaded. The remainder of the mucous surface of the throat, the uvula, and the nasal cavities not being affected by the membrane, which soon loses its transparency, augments notably in thickness, and degenerates into the true diptheritic membrane, of a gray or yellowish color.

"The false membrane is not always continuous, and I have seen several times the tonsils, and the pharynx, in whole or in part, recovered from the membranous deposit while it was progressing upon the soft palate. The membranous fold is easily separated by traction, the use of caustics, and often by Nature, when it appears circumscribed by a red circle. In all these cases it returns again very promptly, but is less thick, and is often reproduced upon a surface, which exhales a fetid and sanious liquid. In some very rare cases the membrane never falls, but is slowly reabsorbed. The voice becomes nasal; the mouth, which rests open, and the nostrils, exude continually an ichorish fluid, which becomes more fetid as the disease progresses, and thickened with the exfoliated shreds of the false membrane. The head, neck, and chest often present a uniform plane, in which the swelling is considerable. Respiration and deglutition are rendered almost impossible, by the increased size of the tonsils and the invasion of the false membrane; the prostration is extreme; the patient is not able to raise his head; the pulse becomes imperceptible, the extremities cold, the intelligence almost always intact, the lips cyanosed, the eyes vitreous, and death comes to terminate the frightful spectacle.

"Such are generally the symptoms when the case terminates fatally. During the first four months of the epidemic, death occurred from the second to the fourth day of the disease, and life was rarely prolonged beyond the sixth. Upon the decline of the epidemic, the progress of the disease was more tardy, and frequently extended to the tenth day."

M. Bretonneau thus describes the specific character of the diptheritic exudation:

"At the beginning of the disease there is perceived a circumscribed redness, which is covered with semi-transparent coagulated mucus. This first layer, thin, supple, and porous, may be still elevated by portions of the unaltered mucous membrane in such a manner as to form vesicles. Often in a few hours the red patches visibly extend step by step, through continuity or contact, in the manner of a liquid which is poured out on a plane surface, or which runs by striæ into one channel. The concretion becomes opaque, white, and thick; it assumes a membranous consistence. At this period it is easily detached, and adheres to the mucous membrane only by very delicate prolongations of a concrete matter, which penetrates into the municiparous follicles. The surface which it covers is usually of a slightly red tint, dotted with a deeper red; this tint is more vivid at the periphery of the patches. If the false membrane be detached, and leave exposed the mucous surface, the redness which appeared subdued under the concretion reappears, blood transudes through the deep red points, the concretion reappears, and becomes more and more adherent upon the points first invaded. It often acquires a thickness of several lines, and passes from a yellowish-white to a grayish and to a black color. At the same time the blood transudes with more facility, and constitutes those *stillicidia* which have been generally remarked by authors.

"Now, the alteration of the organic surfaces is more apparent than at the beginning; often portions of concrete matter are effused into the substances itself of the mucous membrane; there is observed also a slight erosion, and sometimes echymosis in points, which, by their situation, are exposed to friction, or from which the avulsion of the false membranes has been attempted. It is especially about this time that the concretions which have become putrid give out infectious matter. If the concretions are circumscribed, the edematous swelling of the cellular tissue immediately around makes the former appear depressed, and, judging from this appearance only, we might be tempted to believe that we had under observation a foul ulcer with considerable loss of substance.

"If, on the contrary, they are extended over considerable surface, they become partially detached, and hang in shreds more or less putrefied, and simulate the last stage of spachelus; but when we open the body of those who, several days sick, have succumbed to tracheal diptheritis, we shall find in the air-passages all the shades of this inflammation from its first degree, as shown in the portions just invaded, up to that which has, by its deceptive appearance, led us for a moment to dread the supervention of gangrene."

M. Empis regards the commencement of the false membrane as a process of coagulation, which takes place by a precipitation of fibrin, independently of any agency of the living tissue. This is to be seen most distinctly in the air-passages, particularly in the larynx and trachea, in which the tubular cast is seldom adherent, and is commonly much smaller than the cavity it occupies; its external surface, therefore, being separated by a considerable interval from the mucous membrane. That coagulation is not occasioned by the influence or action of the mucous membrane is regarded by M. Empis as proved by his experience in cases where tracheotomy has been performed:

"At the end of a few hours after the operation, whatever care might be taken to clear the canula, the instrument was seen to be lined with a layer of whitish concretions, the thickness of which continually increased. These concretions were evidently only the result of the coagulation of the liquid by which the sides of the canula were constantly covered.

"The pellicle thus formed," says Dr. Slade, of Boston ("Diptheria; its Nature and Treatment"), "which may be considered as the first degree of false membrane, is thicker at the center than at the circumference, and generally may be lifted up, although in very small pieces, owing to its friability. Beneath this superficial pellicle, according to M. Empis, there is still an exudation of sero-mucous matter, which gradually coalesces with the pellicle already formed, thus producing a false membrane several lines in thickness, and adhering to the subjacent surface very closely.

"In many cases the membrane thus formed appears to remain for some time stationary, and then, sooner or later, it takes on an increase in thickness as well as in extent of surface. The secretion of sanious fluid which imbues and softens the concretions is also increased, becomes very dark-colored, and exhales a fetid odor, similar to that of gangrene. This especially applies to the deeper portions of the fauces, to the vulva, and to the anterior parts of the vagina."

M. Empis remarks, with regard to the cicatrization of the membrane:

"We never see the membrane disappear all at once, leaving in its place a cicatrized surface, as is the case with an ordinary eschar, but it is by a gradual process that the pellicle diminishes in thickness, in proportion as the edges of the abraded surface cicatrize. If, however, we modify the secreting surface by an energetic local treatment, we can cause the complete disappearance of the membrane, leaving nothing beneath but a granulating surface of a healthy character."

Any portion of the external surface of the body where the epidermis is absent, and also the surfaces of ulcers and wounds, may become affected with diptheritic exudation as well as the mucous membrane. In some cutaneous affections which have prevailed epidemically, and especially in France, the "cutaneous diptheria" has been a prominent feature. Blistered surfaces, leech bites, excoriations of any part, when the disease prevails epidemically, are liable to become the seat of diptheritic inflammation; and the external manifestation of the diptheritic poison is said to be attended with results quite as disastrous as are its depositions on the mucous membrane.

"When a wound is attacked by diptheritic inflammation," says Dr. Slade, "it becomes painful, fetid, and discolored; serosity pours from it in abundance, and a gray, soft coating soon covers it with a layer of increasing thickness; the edges swell and become violet. The wound remains often obstinately stationary for months; sometimes it spreads, then around it an erysipelatous blush is seen; pustules form, become confluent, burst, and leave apparent a diptheritic patch, which spreads even from the head to the loins.

"A curious fact which has been observed as regards the seat of diptheritic exudation is, that although it is found equally in the mouth, on the soft palate, the tonsils, the pharynx, the nasal fossæ, the larynx, trachea, and even in the bronchial tubes, on the conjunctiva, the vulva and anus, and upon the skin, it is not found upon those portions which are removed from the contact of the air; these seem refractory to the extension of the disease. M. Empis remarks that he never saw true diptheria extend into the esophagus; while, on the contrary, the exudation of certain apthous affections shows a great tendency to spread into the esophagus, but never into the respiratory organs. The atmosphere would thus certainly seem to exert an influence in promoting diptheritic inflammation."

I must reason a little differently from the authors just quoted. I do not think the facts prove that "atmospheric air promotes diptheritic inflammation," but that after the diptheritic excretion has taken place, the influence of the atmosphere will favor the concretion of the fibrinous material exuded into the dense coating which constitutes the false membrane. And this explanation is corroborated, if not absolutely demonstrated, by the fact that the membrane is not unfrequently formed in the bowels, bladder, and uterus, as we have already seen.

M. Empis, who has examined the false membrane microscopically, declares that it is impossible to draw any distinction, founded on microscopic investigation, between the exudation of diptheria and that of a blistered surface, or that which occurs in the throat affection of malignant scarlet fever.

HISTORY OF DIPTHERIA.

The first distinct description of a form of malignant sore throat is found in the writings of Aretæus, who lived about the time of Galen, under the name of Egyptian or Syrian ulcer. Macrobius mentions a similar disease which prevailed in Rome, A.D. 380; and "there is reason to suppose," says Dr. Greenhow, "that we can trace back the history of this affection to a period almost cotemporary with Homer." In 1337 a fatal epidemic of sore throat occurred in Holland. In 1576 it prevailed epidemically in Paris. In 1618–19 it destroyed five thousand victims in Naples; and about this period it prevailed as an epidemic in Spain for forty years. In 1636 it prevailed at Kingston, Jamaica; in 1736 it appeared in Boston, and in 1743 it reappeared in Paris, where it continued until 1748. In 1749 it appeared at Cremona and in England. In 1770 it was first noticed in New York and described by Dr. Samuel Bard.

Dr. Winne, in the paper heretofore mentioned, presents a rapid sketch of the most important historical data, from which I extract:

"It was not, however, until its appearance at Tours in 1818, that it assumed the name of Diptherite, by which it is generally recognized in England and the United States, at the hands of M. Bretonneau, whose investigations have largely contributed to the present fund of knowledge on this subject, and to whom the first connected and practical researches are due. Diptherite made its first appearance at Tours in 1818, in the barracks of the soldiers, in the rear of the legion of La Vendée, and from thence spread to the surrounding quarters. The attack among the soldiers was usually a gingival diptheria, but as it spread into the city the larynx became the seat of the disease, and the gums were not largely affected. From Tours the disease slowly spread to La Ferriére, which it reached in 1824, where, out of two hundred and fifty inhabitants, twenty-one were attacked and eight died. In 1825 the communes north of Orleans were attacked; and in 1828 those south of Orleans suffered from this disease.

"In 1821 M. Bretonneau presented a memoir to the Academy of Medicine, at Paris, on diptheria, as it had prevailed at Tours, which was followed by several others in subsequent years. The whole of his laborious and exact researches were finally given to the world in his treatise entitled, '*Des inflammations speciales du tissu muqueux et en particulier de la diptherite, ou inflammation pelliculaire.*" From the period of its outbreak at Tours, diptheria appears to have seldom or never been absent from one or the other of the departments of France, pursuing a very erratic course, both as to its mode of visitation and the intensity of its attacks, so that the annual reports of the French Academy of Medicine on prevailing epidemics seldom fail to note its existence in some portions of the empire. The visitations, however, which have produced the greatest alarm, not only on account of their severity, but also because of the respectability of the victims, were those of Paris and Boulogne in 1855. The disease at Paris attacked both rich and poor, and while it carried off a large number of children, proved fatal to many adults, more especially those who were often in attendance upon the sick. Among these was the eminent medical writer, Valleix. That, however, at Boulogne was not only the gravest, but of the longest duration, continuing from January, 1855, to March, 1857. During this period it caused 366 deaths, of which 341 were of children under ten years of age. In this epidemic, as in that of Paris, no condition was spared; and, indeed, the attack seemed to fall with the greatest severity upon the children of the wealthy English residents, who, from their more favorable hygienic position, might be supposed to enjoy a comparative immunity from epidemic disease.

"Nor does its fatality appear to have been diminished in subsequent years, for in the report for 1858, read by Trousseau, 22d November, 1859, it is stated that diptheria prevailed in 31 departments, and attacked 1,568 adults and 7,474 children; of these, 165 adults and 3,384 children died.

"In England, the disease first presented itself in the south-eastern counties nearly opposite Boulogne, in the early part of 1857; and traveling from station to station, visited especially the ill-drained and marshy districts, and the neglected and unhealthy localities in towns. Some of the first cases occurred in the practice of Mr. Rigden, of Canterbury, at the beginning of the year. He describes 'seven cases of diptheritic inflammation of the fauces and tonsils, attended with considerable fever, depression and swelling of the tonsils, the fauces and part of the mouth being covered with a pasty lymph.' From this point it gradually diffused itself through the eastern counties, fastening especially upon the marshy districts, in which the attacks were numerous, although the mortality was not in proportion to the number of cases. During the winter months of 1857 it had largely diffused itself through the county of Essex, causing eight out of twenty deaths, and enhancing the rate of mortality in Suffolk and Norfolk in the proportion of three to one. The disease appeared to lull during the summer, but in the autumn of 1858 it largely extended its boundaries, and became quite prevalent in the north midland counties. The county of Lincolnshire appeared to suffer more severely than any other in England, no less than eighty-two deaths being attributed to this cause. In the northwestern counties it prevailed in conjunction with whooping-cough, and in Nantwich caused thirteen out of fifty-nine deaths. It was observed at Wigan, Liverpool, and Hulme, as well as at Rosendale, in which latter place sixteen out of sixty-eight deaths were attributed to its influence.

"Diptheria prevailed at Lima, South America, in 1855, and again in 1858, and is very well described in the concise account given by Dr. Odriazala, a Spanish physician, resident at Lima. In 1855 it appeared in California, and prevailed extensively not only in San Francisco and Sacramento, but likewise in the various mining districts throughout the State. In Placer County it was quite prevalent, but among the districts which suffered most was that of Sonora. The number of cases was very numerous, and the deaths in the aggregate large, but there is no means of determining the relative proportion which they bore to the number affected. Dr. Blake states that at Cache Creek, about twenty miles from Sacramento, the children during 1855 and 1857 were almost decimated by this disease. At Cache Creek it was principally during the spring and summer months that the disease showed itself; and Dr. Bynum, who had attended nearly two hundred cases, states that the affection always appeared more virulent after the prevalence of a north wind, which is a dry and cold one.

"In regard to the conditions under which it appeared, Dr. Blake says it is usually stated that 'it generally prevails in low situations, and to a certain extent this is true; although the most fatal epidemic of the disease that came under my observation was at a mining village called Dutch Flat, situated in a hollow surrounded by hills, about 4,000 feet above the sea. There were thirteen children in the village, all of whom were attacked, and four died. At Grass Valley, which is similarly situated at an altitude of 2,300 feet, the number of cases was great, and the mortality considerable. It was chiefly, however, in the Sacramento valleys and in the valleys of the coast range that the disease was most prevalent.' The disease again renewed its attack in 1858, and is accurately described by Dr. Fourgeaud, in a 'Concise and Critical Essay on the late Pseudo-Membranous Sore Throat of California.'

"The most alarming as well as the most fatal outbreak of the disease in the United States occurred in Albany, in 1858. The first case occurred in the south part of the city, on the 2d of April of that year; the second on the 20th of April, in the same section of the town. From this time it continued to increase in numbers and severity. During the twelve months in which it reigned as an epidemic it attacked about two thousand persons, and caused one hundred and ninety-seven deaths, of which but three were adults.

"The first death from diptheria reported from the office of the City Inspector, in New York, occurred on the 20th of February, 1859, in the practice of Dr. Maxwell; the residence of the child, who was three and a half years old, was in 38th Street, near 5th Avenue. The second death occurred at Manhattanville, on the 25th of February; on the same day, a third fatal case was reported from Stanton Street. On the 5th of March, the fourth case was reported from Vesey Street; on the 10th of March, the fifth from the lower end of 28th Street; on the 23d of March, the sixth from Grand Street, near the East River; and on the 28th of March, the seventh from Varick Street. During the month of April three deaths were reported; in May, three; in June, two; in July, two; in August, four; in September, five; in October, nine; in November, seven; and in December, ten. The whole number of deaths for 1859 was 53, of which 30 were males and 23 females. During the year 1860, the number of fatal cases considerably increased, and the prevalence of the disease as reported at the various Dispensaries was largely augmented. From the 1st to 28th January, 1860, 14 deaths were reported by the City Inspector. For the week ending February 4th, 10 deaths; for that ending the 11th, 12 deaths; week ending 18th, 10 deaths; for week ending 25th, 14 deaths; for week ending 3d March, 19 deaths; for week ending 10th, 9 deaths; for week ending 17th, 13 deaths. The whole number of deaths from diptheria in 1860 was 422.

"Previous to the report of the cases above alluded to, some deaths from diptheria were returned to the City Inspector, but were reported under the head of croup. The number included in this category it is not possible to determine, but it may be fairly inferred that they were not numerous. During the latter part of 1858 and the early part of 1859, a remarkable tendency to affections of the mucous membranes, especially of the throat, was observed, and this became so general as to constitute an important element in the medical man's daily practice. Nor was this confined to any particular part of the city, or class of persons, but seemed to pervade alike the habitations of the opulent, and the confined, ill-ventilated apartments of the poor. As yet, however, no diptheria had been observed, and it was not until about the month of March that medical practitioners here and there, especially among the poor, observed a thin pellicular covering over the tonsils, interspersed here and there with white star-like specks, which gradually expanded in size, and in severe cases came to cover the whole of the tonsils, and extend over the other soft parts of the throat into the larynx on the one side and the nares on the other. This film-like substance could be easily removed with the sponge in its earlier stages, but became dense and closely adherent as the disease progressed.

"Reports of a similar disease have been received from every part of the United States; and in many of the larger places, as Boston, Providence, Philadelphia, Baltimore, Richmond, New Orleans, Cincinnati, Louisville, and St. Louis, as well as in the rural districts, well-marked cases of diptheria have been observed, and in each the bills of mortality have been increased to a greater or less extent through its agency. Although the means of tracing the progress of this disease through the United States do not exist, yet a sufficient number of facts is known to establish that it has not as in England, and to some extent in France, pursued a progressive line of march, but has presented itself here and there in the most erratic manner, and without the general and wide-spread disposition to affections of the mucous membranes which everywhere prevailed, and for the most part still continues."

In his account of the "sweating sickness" in England, in 1517, Hecker says ("Epidemics of the Middle

Ages"): "In January of that year there appeared in Holland another disease which, from its dangerous and inexplicable symptoms, spread fear and horror around. It was a malignant and infectious inflammation of the throat, so rapid in its course, that, unless assistance was procured within eight hours, the patient was past all hope of recovery before the close of the day. Sudden pains in the throat and violent oppression of the chest, especially in the region of the heart, threatened suffocation, and at length actually produced it. During the paroxysms the muscles of the throat and chest were seized with violent spasms, and there were but short intervals of alleviation before a repetition of such seizures terminated in death. Unattended by any premonitory symptoms, the disease began with a severe catarrhal affection of the chest, which speedily advanced to inflammation of the air-passages. In Basle, within eight months, it destroyed 2,000 people."

In the year 1736, Dr. Douglass, of Boston, published an account of the first appearance of a "sore throat distemper" in this country. The epidemic which he describes was very malignant, and was attended with "erysipelatous appearances and highly putrid symptoms."

Under date of October 1, 1753, Mr. Cadwallader Colden addressed a letter to Dr. Fothergill concerning the "throat distemper," which was published in the first volume of "Medical Observations and Inquiries," London. Mr. Colden says:

"The first appearance of the throat distemper was at Kingston, an inland town in New England, about 1735. It spread from there, and spread gradually westward, so that it did not reach Hudson's river till nearly two years afterward. It continued on the east side of Hudson's river before it passed to the westward, and appeared first in those places to which the people of New England resorted for trade, and in the places through which they traveled. It continued to move westwardly, till I believe it has at last spread over all the British colonies on the continent. Children and young people were only subject to it, with a few exceptions of some above twenty or thirty, and a very few old people who died of it. The poorer sort of people were more liable to have the disease than those who lived well with all the conveniences of life, and it has been more fatal in the country than in great towns.

"In some families it passed like a plague through all their children; in others, only one or two were seized with it. Ever since it came into the part of the country where I live (now about fourteen years), it frequently breaks out in different families and places without any previous observable cause, but does not spread as it did at first. It seems as if some seeds, or leaven, or secret cause remains wherever it goes. When the distemper becomes obvious, it has the common symptoms attending a fever, except that a nausea or vomiting is seldom observed to accompany it. The disease is not often attended with that loss of strength that is usual in other fevers; so that many have not been confined to their beds, but have walked about the room till within an hour or two of their death; and it has often appeared no way dangerous to the attendants, till the sick were on their last agony. Some died on the fourth or fifth day, others on the fourteenth or fifteenth day, or even later. When this disease first appeared, it was treated with the usual evacuations in a common engina, and few escaped. In many families, who had a great many children, all died; no plague was more destructive."

One source of the fatality of the epidemic described by Mr. Colden is indicated in that significant line, "it was treated with *evacuants*, and few escaped." The whole history of all malignant epidemics shows that the depleting practice of the physicians has caused more deaths than would have occurred had the disease been left to itself, and the powers of life to their own unaided resources.

A throat-disease, in all essential particulars, prevailed in Sullivan County, N. Y., in November and December, 1861 (principally in the town of Lock Sheldrake and its immediate vicinity), and although there was no dearth of doctors (from three to six consulting together in some of the cases), *every case* proved fatal. Several families lost all of their children. In the family of a Mr. Kyle, of eleven children, nine died. Almost all the deaths in this neighborhood occurred in about forty-eight hours after the first alarming symptoms. A large proportion of these patients were adults.

In the epidemic sore throat which prevailed in New York in 1771, as described by Dr. Bard, the disease was generally confined to children under ten years of age. The symptoms were usually so mild for five or six days as to create little alarm; after which occurred very great and sudden prostration of strength, a peculiar hollow, dry cough, and a remarkable change in the tone of the voice, indicative of ulceration in the laryngeal passage, or a concretion of the exuded lymph.

In the cases described by Dr. Bard, the swelling of the parotid, sublingual, and submaxillary glands, mentioned by other authors as invariably present, was noticed in but few instances.

In the spring of 1860, the disease appeared endemically near New Haven, Conn., and is thus described by Dr. L. N. Beardsley, of Milford—who attended the first fifteen cases—in a communication to the Boston *Medical and Surgical Journal:*

"This disease [diptheria] appeared in an endemic form, and with great mortality, in this vicinity during the months of March and April last. It first made its appearance in Orange, an adjoining town (which is in an elevated situation, and is a remarkably healthy place, with a sparse population), and for a while was confined entirely to the scholars attending a select school in the village. Fourteen out of fifteen of the cases, of those who were first attacked, proved fatal, in periods varying from six to twenty-four days.

"Most persons residing in the district where the disease first appeared, sooner or later had some manifestations of the disease. The period of incubation varied from five to twenty days. The lymphatic glands were in many cases greatly enlarged. The first symptom—and it is one which we have never seen referred to by any writer on the subject—was *pain in the ear.* It was not only pathognomonic but prominent, and almost invariably present in every case that came under our observation, in a day or two before the patient made the least complaint in any other respect, and before the smallest point or concretion of lymphatic exudation could be discovered on the tonsils, or elsewhere."

The language of Dr. Beardsley is a little muddled. To be "pathognomonic," a symptom should be *invariably* present, and not "almost invariably," as our author expresses it.

During the years 1860 and 1861 diptheria has prevailed sporadically or endemically in nearly all sections of the United States, and at this time, so far as I can learn from extensive correspondence with all parts of the country, it seems to be on the increase.

INFECTIOUSNESS.

Is diptheria contagious? This question has been much discussed by medical writers, and, as has been the case with scarlet fever, yellow fever, plague, and some other diseases, the testimony adduced *pro* and *con* seems to be pretty equally balanced. I suspect that the disputants on both sides of the question in issue are partly right and partly wrong. I am of opinion that under certain circumstances all febrile diseases may be contagious. In all fevers there are morbid excretions which, if due attention is not paid to ventilation and cleanliness, may become so accumulated and concentrated, as it were, as to infect other persons, and thus become the cause of a similar disease in them. Much depends, of course, on the greater or less susceptibility of the individual to be affected, and this susceptibility, or non-susceptibility, is nothing more nor less than the grossness or purity of the party exposed.

Several authors, among whom is M. Bretonneau, have maintained that the exuded matter of diptheria possesses a special virulence, and that the disease may be propagated by the application of the secretion [excretion?] from an affected surface to sound parts, like small-pox; and he contended that, like syphilis, the disease can only be communicated by contact, thus rendering it technically *infectious*, as is syphilis, instead of both *infectious* and *contagious*, as is small-pox.

M. Bretonneau says: "Innumerable facts have proved that those who attend patients can not contract diptheria unless the diptheritic secretion in the liquid or pulverulent state is placed in contact with the mucous membrane, or with the skin on a point denuded of epidermis, and this application must be immediate.

"The 'Egyptian disease' is not communicated by volatile invisible emanations, susceptible of being dissolved in the air, and of acting at a great distance from their point of origin. It no more possesses this quality than the syphilitic disease. If the liquid which issues from an Egyptian chancre, as visibly as that which proceeds from a venereal chancre, has seemed under certain circumstances to act like some volatile forms of virus, the mistake has arisen from its not having been studied with sufficient attention. The appearance has been taken for the reality."

In support of the doctrine of the infectious nature of diptheria, M. Bretonneau has adduced the following among other cases: In the hospital at Tours, a child affected with diptheria, in a fit of coughing, ejected a portion of the diptheritic matter which lodged upon the aperture of the nostril of M. Herpin, the surgeon, who was at the time sponging the larynx of the patient. This M. Herpin neglected immediately to remove, and the result was a severe diptheritic inflammation which spread over the whole nostril and pharynx, attended with extremely severe constitutional symptoms, with great prostration, and followed by a slow and lingering convalescence of six months' duration. This child, it, is also stated, had transmitted the affection to its nurse.

Dr. Gendron, of Chateau de Loire, having received on his lips portions of diptheritic exudation, expelled by a patient in the act of coughing, was soon affected with a violent laryngeal inflammation.

In 1826 a boy, affected with frost-bites of his foot, had a painful diptheritic inflammation of the great toe, soon after using a bath that had been employed for a diptheritic patient.

M. Bretonneau concludes from his experiments and observations, that the disease can not be occasioned by atmospheric communication, and is not therefore contagious; but that its cause is transmissible by inoculation, and is therefore strictly infectious. And his observations are corroborated by other authors.

But, on the other hand, Prof. Trosseau, who inoculated himself and two of his pupils with diptheritic matter, failed to produce any results whatever; and Dr. Harley, of London, was equally unsuccessful in experiments on domestic animals.

The experience of M. Isambert, of Paris, as related in a communication on the epidemic of malignant sore throat which occurred in Paris in 1855, goes to prove that diptheria is contagious as well as infectious. He says: " Diptheritic affections sometimes appear sporadically; they also often seem to be endemic, as well as epidemic and contagious. As predisposing causes, we may consider that the lymphatic temperament, a feeble constitution, privation, etc., all exert a decided influence. Youth is much more exposed to the disease than any subsequent age. Locality and overcrowding have a positive effect; so also do cold and changeable seasons.

" Epidemic influences are much the most powerful. As to the contagious nature of the disease there can be no doubt, since many physicians have contracted it. The opinion of M. Bretonneau, that diptheria is not transmitted by the atmosphere, but is always the result of inoculation, is altogether too exclusive. With M. Trosseau, we can not reject infection at a distance as one of the means of propagation possessed by diptheria."

That the prevalence of the disease depends quite as much on the condition of the inhabitants as upon the moisture or temperature of the atmosphere, or even upon the vague and indefinite " epidemic influence," so much talked of and so little understood, is shown by the following facts adduced by M. Trosseau : " In the villages of the Loire, remarkable for their salubrity and for their excellent position, I have seen diptheria prevail to a terrible extent, while the villages of Sologne, situated in the midst of marshes, remained exempt; and, again, hamlets bordering on ponds depopulated by the epidemic, while others enjoyed a complete immunity."

And so, too, I have known the most malignant and putrid forms of typhoid fevers and of dysentery, and of erysipelas, and of scarlatina, as well as diptheria, prevail in the most salubrious places in New England and New York, and in as healthful localities, probably, as the sun ever shone upon. And there, as elsewhere, I suspect the essential causes are to be found chiefly in the habits of the people. The epidemic influence is within the vital domain itself, instead of the atmosphere without.

Dr. Samuel Bard, as well as nearly all the writers of the seventeenth century, considers the disease to be infectious. He says: "The disease I have described, appeared to me to be of an infectious nature, and as all infection must be owing to something received into the body, this, therefore, whatever it is, being drawn in by the breath of a healthy child, irritates the glands of the fauces and trachea as it passes by them, and brings about a change in their secretions. The infection, however, did not seem, in the present case, to depend so much on any generally prevailing disposition of the air as upon effluvia received from the breath of infected persons. This will account why the disorder should go through a whole family and not affect the next-door neighbor."

Dr. Ranking, in his late lectures on diptheria, has probably presented the subject of its contagiousness or infectiousness in the true light. He remarks: "My own conviction is, that it is infectious to a limited degree; by which I mean, that when patients are accumulated in small, ill-ventilated rooms, the disease is likely to be communicated; but I do not fear that, like scarlatina or erysipelas, it may be propagated in spite of all sanitary precautions. Still less that the infection can be conveyed by the clothes or persons of those who visit or superintend the patients. That it commonly spreads through the family once invaded is to be attributed, in some degree, to the persistence of the same cause as originated the first case."

Dr. Edward Ballard, of Islington, in an article published in the *Medical Times and Gazette*, July 23, 1857, adduces the following, among other facts, in support of the infectious character of diptheria: "Infectious diseases habitually spread in families they invade. Out of forty-seven families there were only fifteen in which the other members all remained healthy. Of course it may be argued, in opposition, that all the members of a family are equally exposed to the operations of local causes of disease. As a rule, diptheria spread in the houses it invaded, chiefly among those members of the several families who were most closely in communication. In no case where separation from the sick person had been effected early in the disease, have I noticed that it has spread to the separated individuals."

Although we admit that persons affected with diptheria may communicate, under favoring influences, the causes of the disease to others, it must be obvious that whatever local or other causes occasion it in any one member of a family, are also liable to induce it in all the rest, quite independently of contagion or infection. And in point are the results of inquiries instituted in fifty-seven houses where fatal cases have occurred: "In more than half of these houses there was some defect in the sanitary arrangements, or in the surrounding conditions of the patient. In the greater number of the houses thus deficient, the fault was discovered in the state of the drainage."

Per contra we have an equally formidable array of medical authorities who contend that diptheria is not infectious at all, and but feebly if at all contagious.

M. Daviot, who has written a memoir on the disease, declares that he has never met with an instance where it was communicated by personal intercourse; and that neither the attendants nor those who cauterized the throats of affected children contracted the disease. Negative testimony, however, should have but little weight against positive. What one physician has seen is not to be disproved by what another has not observed. A disease may be infectious or contagious, and prevail in different places and under different circumstances in the same place, epidemically, endemically, or sporadically. All persons who are brought in contact with patients affected with small-pox, measles, or syphilis, do not have the disease; and it not unfrequently happens that only one person in a neighborhood will have small-pox, measles, whooping-cough, mumps, etc.

Dr. Crighton, of Edinburgh, treated forty-five cases of diptheria, of which nine resulted fatally; and in reporting the cases he remarks: "In only two cases was there anything like proof of contagion, and, from all that I have seen of diptheria, I believe that, although it would be incorrect to separate it from the list of communicable diseases, yet it is very feebly so compared with many others. I may mention one instance which struck me particularly, where, in a large family of six or seven children, and chiefly under the age of twelve, a child had the disease in a very severe form, and although he was never isolated during the day from the others, but lay on a sofa in a room where I generally found several of them at my visit, they all escaped."

Dr. Monkton, of Kent, England, who has had a large experience in diptheria, reports through the *Medical Times and Gazette* of February 26th, 1857: "No decisive instance of the communicability of the disease has come before me; on the contrary, I have seen it attack individuals only, in a family of liable persons, much more frequently than I think scarlet fever would have done. My own conviction is, that diptheria is epidemic, endemic (*i. e.*, largely affected by locality), and non-contagious, or, if contagious at all, vastly less so than scarlet fever, from which it is very distinct."

Dr. Slade remarks: "Now, although those who favor the idea of contagion find in the phenomena of cutaneous diptheria strong ground for the support of the theory of inoculation, there are facts which would equally seem to oppose it. For example: it has been observed in these epidemics that the false membrane upon the skin not only presents itself in those not previously affected with faucial diptheria, but it not unfrequently attacks remote parts, such as we should suppose were inaccessible to inoculation, as, for example, the folds of the groins in children, and the spaces between the toes. A single well-observed fact of this kind is sufficient to cast a doubt on the theory of inoculation."

On the subject of the communicability of the disease, Dr. Greenhow remarks: "Although I have no proof that diptheria is communicable by means of the exudation, many facts have fallen under my notice which convince me that the disease is in some way or other communicable. I attach little importance to the circumstance that diptheria so often attacks simultaneously, or at short intervals, several members of the same family; such facts may be explained on the supposition that the patients have in such instances been all exposed to one common cause, be it endemic or epidemic. But if, soon after the arrival of a patient from an infected district, diptheria should break out in a place where it did not previously exist, and attack persons who have been in direct communication with the invalid, and especially if it attack only such persons, then have we the strongest presumptive evidence of its being a contagious disease."

The facts already adduced—and a multitude of similar ones could be easily collated—prove to my mind, most clearly, that diptheria originates indigenously, and that it may be communicated, under peculiar circumstances, from one person to another.

CAUSES OF DIPTHERIA.

On no subject is medical literature more crude, vague, unsatisfactory, and irrational than in relation to the causes of disease. And this must ever be the case so long as the medical profession confesses profound ignorance of the essential nature of disease. When this primary problem is solved, when medical men understand what disease is, they will not be long in comprehending the causes which produce it, at least with sufficient accuracy and exactitude for all practical purposes.

For hundreds of years the profession, in its investigations of the causes of disease, the nature of disease, and the remedies for disease, has been pursuing a mere phantom. Medical men have assumed that diseases have specific characters or natures inherent in themselves, and that therefore each must have a specific cause, and require a specific remedy. There can be no greater delusion. And when we reflect for a single moment that disease—that all disease—is the action of the living system to resist poisons, expel impurities, or to repair damages; that it is purely and simply a defensive or remedial struggle—vital action in relation to things abnormal—this whole doctrine of specifics will appear sufficiently absurd.

There are but two sources of disease, aside from mechanical injuries or irritants, and there are, as was explained by Hippocrates nearly three thousand years ago, poisons introduced from without, or impurities generated within. If we inhale miasms or particles of foreign substances which float in the atmosphere, or if we absorb them through the skin, or if we take them into the stomach in the shape of aliments, condiments, or medicines, the blood becomes impure and the capillary vessels obstructed. Or if the waste or effete matters of the system—the ashes or *débris* of the disintegrated tissues—are not properly deterged by the various emunctories, impurities are ingenerated; that is to say, the effete matters which should have been expelled are retained, causing obstructions, and thus becoming the occasions or causes of disease; the disease itself, let it never be forgotten, is the effort of the living system to remove these obstructions.

The particular form of disease, the manner in which the remedial effect, or the process of purification will be manifested, must depend on a variety of circumstances and conditions—the nature and quantity of the obstructing materials, the absolute and relative vigor and integrity of the various depurating organs, the habits of the patient, atmospheric, electrical, thermometrical, and passional influences, etc.

We are taught in medical books that certain diseases have inherent dispositions or tendencies to *impress* or *act upon* particular parts of the system, or to *locate* in certain organs, or to *seat themselves* here or there, or to *run through* the system, etc., etc., all of which vagaries are founded on a false notion of the nature of disease.

Dr. Jacob Bigelow, of Boston, who claims the honor of being the father, or at least one of the fathers of the modern doctrine of "self-limited" diseases, gives us, in a late work ("Nature in Disease") the following lucid exposition of the subject: "By a self-limited disease, I would be understood to express one which *receives limit from its own nature*, and not from foreign influences; one which after it has *obtained a foothold* in the system, can not, in the present state of our knowledge, be *eradicated* or abridged by art."

If disease is really an independent entity, a thing, a foreign substance, a something outside of the living organism, a being or creature analogous to a ghost or goblin, imp or sprite, fiend or demon, spook or spirit, such reasoning might be the very quintessence of medical philosophy. But if disease is in fact nothing of the sort; if it be the exact contrary, if the disease and the *vis medicatrix naturæ* be one and the same thing, as I hold to be true and demonstrable, then, in the light of this truth, nothing can be more ridiculously nonsensical than Dr. Bigelow's explanation of a self-limited disease. Disease is represented, by the learned Doctor, as a creature or thing which has obtained a "foothold" in the system, and after having *established itself* in the vital domain, it then *ordains for itself* a law of limitation, and receives limits from its own nature. Is there not something incomprehensibly queer in the idea of a disease taking forcible possession of a living body, then *dictating to itself* laws and limits, affixing to itself boundaries of time and space, selecting the place of its abode, and determining just how long it will stay or go, or exist, or remain, or run, or be seated, or where, and when, and how, and why it will consent to be unseated, and utterly refusing to be "eradicated or abridged" by the art of dealing out all the drugs of the apothecary shop?

The error lies further back. It consists in mistaking the relations of living and dead matter. Medical books and schools teach that the causes of disease act or make impressions on the living organism, and that diseases do the same, and that remedial agents do the same. The reverse of all this is the truth, as taught in the Book of Nature and in the School of the Universe.

As a general statement, poisons, impurities, or organic lesions are the direct or immediate causes of all diseases, and unphysiological habits or conditions are the causes of these causes—the remote or predisposing causes of disease. But it is very difficult to detect the nature or properties of those poisons or impurities—morbific materials—which result from the changes, transformations, and decompositions of organic matter. They elude all the art of the chemist, all the skill of the anatomist, and defy the vision of the microscopist. An almost inappreciable quantity of variolous matter, for example, applied to any part of the living body denuded of its cuticle, will occasion a violent fever attended with a pustular exanthema over the whole surface; yet the analytical chemist has never been able to ascertain the constituent elements of that virus. And, indeed, chemistry never has been and never will be able to determine the exact composition of any organic product, whether it be food, tissue, effete matter, secretion, excretion, poison, or virus, for the reason that, in the process of analysis, some of the elements are changed or lost. Chemistry can determine what remains as the result of the analysis, and this is all. And when chemists undertake to tell us what food is, what disease is, what vitality is, what living tissue is, or what remedies are, by a process of chemical analysis, they are entirely out of their proper element. These problems are all to be determined by physiological laws, not by chemical decompositions; by the instincts of the living organism, and not by the constitution of dead matter.

In croup, and in diptheria, and in certain other morbid conditions of the system, in the process of depurating the system of its virus or impurities, the fibrinous or albuminous elements of the blood are exuded upon the skin or upon the mucous surfaces; in cholera, the serum of the blood is poured into the intestinal tube; in eruptive fevers, some morbific material is expelled through the skin; in diabetis, a saccharine element is deterged copiously through the kidneys; in diarrhea, fecal matters are dejected by the bowels; in consumption, tuberculous matter is expelled through the pulmonary structure; in cholera morbus, vitiated and acrid bile is excreted from the liver; in simple fevers, effete matters of various kinds are determined with more or less force to one or more of the depurating organs, etc. In all of these cases—and the principle applies to all diseases—the form of the disease and the nature of the material excreted or expelled, depends on—1. The force and direction of the remedial effort. 2. The organ or structure through which the purifying process chiefly takes place. 3. The condition of the whole mass of blood at the time. 4. The quantity of morbific material to be eliminated; and 5. The external influences operating, so to speak, at the time, as temperature, humidity, etc.

Much has been said and written on the influence which meteorological and cosmic conditions exert in the production of cholera, diptheria, and other pestilences; but the whole subject is scarcely better understood now than it was before the sciences of meteorology and cosmogony were heard of. Bretonneau entertains the notion, that diptheria could only be developed in a damp atmosphere. But in the recent epidemics of France and England, the disease has prevailed in high and dry situations. And in this country

I am not aware of any facts which tend to prove that it is more prevalent or more severe in damp localities than in dry. In California, noted for its very dry, summer atmosphere, according to Dr. Wooster, the disease has been very prevalent and very fatal. Dr. Wooster states, in a monograph on diptheria:

"In our climate, the air in summer becomes so dry, that if an ordinary soft wooden pail or bucket be half filled with water, and set in the sun in the open air for six hours, and then two quarts of water be added, it will leak through the joints of the shrunken staves, above the surface of the first portion of water. A miner uses a bucket to bail water from a hole all the forenoon, and, although it is perfectly saturated with water, yet if he leaves it in the sun while he goes to his dinner, when he returns it will often fall to pieces as he attempts to take it up.

"This is the kind of air in which the disease has occurred with unequaled fatality in this State. In this city I can not ascertain that a case has occurred in that part of the town built over or near the waters of the bay, or on the salt marshes near it. But I have seen cases in the high part of the city, and on bluff headlands extending into the bay, points that from their elevation and constant exposure to a strong breeze would be thought inaccessible by any morbid [morbific?] effluvia."

It should be considered here, that the inhabitants of high and dry situations may live in the line of the currents of wind which convey the miasms of the low and wet localities, and hence, although their situations are in themselves perfectly salubrious, the people residing there may be really more exposed to miasmatic diseases than are the people who dwell in the more humid atmosphere of the lowlands, out of the direction of the prevailing winds.

According to Mr. Ernest Hart, the diptheria has appeared in France and in England with no regard whatever to any recognized climatic or meteorological laws. It has visited the open hamlets of the rural departments, and the crowded courts of the great cities; it has prevailed at the sea-side; in the heat of summer; during the cold of winter; in marshy, ill-drained localities; in dry and elevated regions; in ill-ventilated barracks, and in the open country; in dry places; in damp places; in the low valleys, and on the high mountains.

There is truth—practical truth—in the following paragraph: " Zymotic in its nature, it tends to fasten upon whomsoever is debilitated by previous disease, or by a constitution naturally feeble and artificially effeminized, or where vitality is lowered by the depressing influences of luxury, indolence, and inactivity; and the habitual defiance of physical and hygienic laws, which is so frequent an element in fashionable life. Hence individual cases come into play, and introduce this associate of the poor into the palaces and mansions of the great, which they so often fringe. Diptheria finds there its victims pale and anemic, or grossly sanguineous and unhealthily excited."

"Grossly sanguineous!" Bad blood is the essential condition of all putrescent, pestilential, and malignant diseases, and gross living is the essential cause of bad blood. And when we investigate the etiology of diptheria to its starting-point, I suspect we shall find that impure or indigestible food, with inattention to personal cleanliness—the chief sources of impure blood and foul secretions—are the essential causes of diptheria.

Nor can I forbear alluding in this place to what I can not help regarding as standing at the very head of the "specific" causes of this disease—swine raising and pork diet. That the flesh and grease of that filthy scavenger, the hog, in the form of pork, ham, sausages, lard, etc., constitute a most impure and blood-poisoning aliment, I believe no intelligent physiologist will deny. And that a sty-fed hog is a diseased carcass, is evident to all pure senses. That pork and scrofula stand to each other in the relation of cause and effect, has been proverbial among observing men for centuries. Yet all over this Christian land some form of sty-fed and sty-fattened hog-food is one of the most common, most cherished, and most relished dishes to be found on the tables of the rich or poor; while in an ordinary hotel, boarding-house, or restaurant, or even in a private family, but few articles of food can be found not attainted with some part or portion of the tissue or adipose matter derived from this disgusting animal.

Within a few months I have visited and lectured in different States—Maine, Massachusetts, Illinois, Indiana, Iowa, Ohio, and in the District of Columbia—and in all places I inquired particularly as to the prevalence of diptheria, and also as to the dietetic habits of the people, especially with regard to pork-eating. In each State which I visited I heard of places where the disease had been very prevalent and very fatal, and in all of these places swine-food was employed very freely, as was swine-grease, as shortening for pastry, cakes, biscuit, and even bread.

But pork-breeding as well as pork-eating conduces to this as well as to other foul and malignant epidemics. And I am of opinion that all of the contagious diseases in the world originate from slaughter-houses, hog-pens, distilleries, barn-yards, stables and henneries, provision dépôts, etc., where animal offal and excrements accumulate, and where animal matter is constantly undergoing decomposition and putrefaction, thus loading the atmosphere with miasms and impurities. If the people would all become vegetarians, there would be, in my opinion, an end at once of such diseases as eruptive fevers, and of contagious diseases of every sort.

Persons who do not eat pork, but who dwell in close proximity to piggeries, may become infected with the seeds of diptheria or some other foul disease. The very atmosphere is poisoned with the effluvia which constantly emanates from the lungs and skin and excrement of the animal, so that one who abhors the unclean aliment may be destroyed by inhaling the miasms which the noxious animal generates, as one who does not smoke cigars nor chew tobacco may be nauseated and sickened by the breath and atmosphere which are rendered poisonous and pestilential by those who do smoke and chew.

MORTALITY OF DIPTHERIA.

The mortuary statistics of no disease present greater diversity of results than those of diptheria. This may be accounted for, in part, by the great diversity of circumstances under which the disease prevails, and the different habits and constitutions of the persons who are the subjects of it. Much, however, is due to the course of medical treatment adopted. And I fear that a careful investigation of this branch of our subject would disclose another evidence of the truth of the saying, "Just in the ratio that doctors and drugs have increased, diseases and deaths have multiplied." So far as I have been able to collect statistics bearing upon this point, I find no exception to the general rule, that the mortality of diptheria is everywhere in proportion to the potency of the drug-medication.

As is the case with malignant scarlet fever, croup, typhoid pneumonia, and other diseases of low diathesis, many cases will *bear* drug-treatment, bleeding, blistering, etc., which are not benefited by them, while in the most severe cases their administration is almost certain death. Physicians have often noticed and recorded the fact that, in the treatment of *scarlatina maligna*, a single dose of castor-oil, or a moderate bleeding, has destroyed the life of a patient in a few hours. But in the milder form of this disease—*scarlatina simplex*—the patient will bear repeated bleedings and purgatives, and survive both the disease and the medication. These remarks apply with equal force to diptheria. The discordant methods of practice to which different physicians have resorted, and the disagreements of medical men of extensive observation and large experience, with regard to what is useful or injurious in its treatment—some authors condemning as aggravating the disease and endangering the life of the patient the very remedies and measures which others rely upon as essential to the cure—render an estimate of the necessary or even actual mortality of the disease itself, independent of treatment, a very difficult matter to determine. In their reports of cases, medical men almost universally assume that all patients who recover are indebted for their lives to the drugs and doses administered by the physician. But this is by no means a logical conclusion. It may be that patients recover in *spite* of the medicine, rather than with its assistance. And this I believe to be the general if not the universal rule, so far as drug-medication is concerned. Medical reasoning is unlike all other reasoning. It seems to disregard all the ordinary rules of logic; and medical men have a method of ratiocination peculiarly professional and exclusively medical. They claim that their remedies have a power over the vital functions, and that they are capable of controlling morbid actions; and when a patient recovers, the remedies employed are accredited with the cure. But, suppose the patient dies? what then? Do these medical logicians charge the *killing* to the medicine? Never. The patient dies in spite of it. This is not legitimate logic. It is just as rational to assume that when the patient dies, the medicine *kills* him, as to assume that when the patient recovers, the medicine *cures* him. Indeed, this is the more reasonable assumption, because the relation of the drug to the living organism is that of a poison; its tendency is to kill.

These views are corroborated by the observations which I have had the opportunity to make in various parts of the country, with regard to the results of Hygienic and of drug treatment. In all cases where little or no medicine was employed, the per centage of deaths was very small. In all cases where active drug treatment was resorted to, the proportion of deaths was large. In some places where the disease has prevailed endemically, every case has been treated hygienically, without the loss of a single patient; while in other places, all the cases have been treated with the ordinary drug remedies, and all have died.

These extreme results surely mean something; and before we can determine diptheria to be a dangerous disease, *per se*, we must understand the effects of the various methods of treatment brought to bear upon it, or upon the patient.

While lecturing in Boston, recently, I met with Dr. Prescott, of Farmington, Me., who, after an experience of forty years of drug-medication, came to the same conclusion that Prof. Jos. M. Smith, of the New York College of Physicians and Surgeons, has recently arrived at, that "drugs do not cure disease; disease is always cured by the *vis medicatrix naturæ*;" and to the conclusion that Prof. Alonzo Clark, of the same school, not long since announced to the medical class, viz.: "All of our medicines are poisonous, and, as a consequence, every dose diminishes the patient's vitality;" and who, acting in accordance with his honest convictions of truth and duty, renounced drug-medication and adopted the Hygienic, which system he has advocated and practiced ever since.

Dr. Prescott informed me that in one of the adjoining towns, of thirty-five cases of diptheria, thirty terminated fatally. Of course these cases were treated with the ordinary remedies. Dr. Prescott has himself treated several cases hygienically, and has lost but one. In October last, I visited and lectured in Augusta and in Vassalboro', Me., and there learned the statistics of the mortality of the disease as it had prevailed in various parts of the State. The whole number of deaths was about one half of all the cases, although the mortality varied in different places from twenty to seventy-five per cent. In one town—Canton, if I recollect—of twenty cases, sixteen were fatal. In a few instances the friends of the patients had managed the cases as well as they could with "water-treatment," and of these none were lost. During a lecturing excursion in December and January last, in some parts of Ohio, Indiana, Illinois, and Iowa, I heard of many places in which the disease had appeared, and the average mortality was nearly the same as in New England. In some places nearly all the patients died, and in other places a very small proportion of the cases were fatal. In Sullivan County, N. Y., where the disease prevailed endemically, as I have already remarked, all the cases terminated fatally. And in Orange, Conn., according to the report of the attending physician—Dr. Beardsley, of Milford—of the fifteen cases which occurred, fourteen were fatal. And it may not be irrelevant to remark that, being on a professional visit in the neighborhood of Orange, soon after the occurrence of the endemic in the place, I was informed that the fifteenth case—the only patient who survived—removed from the place, *and from the Doctor*, so soon as the disease attacked him. What relation this circumstance had to the recovery is, of course, a legitimate question for differences of opinion.

Dr. Fougeaud, in a monograph on a terrible epidemic which prevailed in San Francisco, and in other towns in California in the autumn of 1856, having all the characters of pharyngeal diptheria, states that the mortality was appalling. " Few children attacked by it recovered."

In the city of New York no record of death of diptheria is to be found anterior to 1857, and in that year only two cases are recorded. In 1858 only five cases were reported to the Inspector's office. In 1859 the cases had increased to fifty-three, and in 1860 four hundred and twenty-two deaths were reported; since which the mortality has averaged about fifteen per week.

During the last two years, newspaper accounts have come to us from more than a hundred places in the United States, stating that families had lost three, four, five, six, seven, and even more members of the disease, and in not a few instances all of the members of the family have died. And during this time I have received some hundreds of communications, asking for information respecting the proper Hygienic treatment of the disease, of whose nature and contents the following extracts will serve as samples:

LACON, ILLINOIS, *Nov.* 28, 1860.

R. T. TRALL, M.D.—*Dear Sir:* Diptheria is raging all around us. All ages are having it. With children it is very fatal—all, in fact, die. The people in this community are most completely blinded by drugopathy, and are doctored in what seems to my common sense to be the most outrageous manner—blisters, turpentine, quinine, brandy, beef-tea, etc., etc. I have read what you have written on the subject in the *Water-Cure Journal*. Myself and wife have been among it constantly, and so far we keep well, as do our five children. S. S. H.

SANDWICH, MASS., *Jan.* 16, 1861.

DR. TRALL—*Dear Sir:* The diptheria has made its appearance on all sides of us, but has not yet entered our village, though a widow lady, who moved from our town some months since, has lately brought here for interment the remains of four children, the victims of the new disease. Under the circumstances, I very naturally feel a little anxious for my only child, a boy of nine years of age, for, should he be attacked with the disease, I should be very loth to subject him to the tender mercies of the regular system of practice. I wish to treat him hydropathically should occasion require, and to that end I desire such information and advice as you can give me. I should perhaps state that I am a humble mechanic, yet I trust with sufficient common sense to understand the rationale of water-cure; for, guided by the instructions of your "Hydropathic Encyclopedia," I have been able for several years to dispense entirely with drug-medicines, and with the advice of physicians.

The reports of physicians who claim remarkable success in the treatment of diptheria should be taken with some grains of allowance, for it not unfrequently happens that when throat-affections are numerous, with occasional or sporadic cases of membranous exudation, all will be grouped together under the head of diptheria; and for this reason it is exceedingly difficult, in many cases, to judge of the merits or demerits of the treatment adopted. Some physicians have claimed unusual skill, or a superior *methodus medendi*, because they have lost less than ten per cent. of the cases they have treated; but it is not certain that, had there been no physician in the case, even the ten per cent. of mortality might not have been diminished.

Professor Alonzo Clark, in his recent lectures on diptheria, in relation to its mortality, remarks:

"We now turn to the mortality of diptheria. This is a complex problem. Its fatality in its different occurrences and in different places varies more than that from any other disease I can name to you. In certain families, schools, and villages, the deaths among those attacked are more numerous than from any other epidemic affection. The proportion is not less than that from membranous croup or tuberculous meningitis.

Under other circumstances, no more than one case in forty or fifty proves fatal. But this remarkable difference, while there is really the membranous inflammation, is less embarrassing than the fact that many physicians, who have reported their observations, have not separated their cases of simple tonsillar inflammation without any pellicular exudation, from those of true diptheria; but on the contrary, professing to regard the two affections as arising from the same cause, since they prevail at the same time, have grouped them together, and so have greatly reduced their proportion of mortality. I have already said that there is but one way of treating this matter fairly. It is to make the membrane the basis of classification. This will separate diptheria from everything but the true croup; and the marks soon to be indicated will commonly be sufficient to establish the distinction between these. Indeed, the epidemic character of one and the sporadic occurrence of the other will, of itself, be enough, at least in the great majority of instances. It may be said that this distinction between simple and diptheritic sore throat is not scientific. It may be so; but it is better than scientific, it is practical. The aim and end of science is the improvement of man's condition. If this improvement can be but attained by dividing into two groups what Nature allows us to class as one, no rule of sound reason can forbid the separation. Besides, science does require us to make distinctions when there are differences. And here we have the broad difference that one disease is ephemeral, with a uniform tendency to recovery; the other is often terribly fatal, or it is liable to a long train of sequences of a serious, if not of an alarming character.

"An example will enable you to understand how far

I would have you carry the distinction. Five children in one family had sore throat, all occurring within the same week. The same cause had probably operated on all, but the effects resulting from that cause were different in the different children, and the difference, no matter how produced, was cardinal so far as the safety of these patients was concerned. Three of them had active fever, flushed face, pain in swallowing, the tonsils were swollen and red, and on the anterior and inner surface of each were half a dozen or more yellowish white spots. The matter constituting these spots was an opaque concretion from the follicles of these organs, and each mass was nearly globular in shape, and was embedded in the tonsil. To a person not familiar with their appearance, they might have been mistaken for false membrane; but they were rounded, and the false membrane is flat; they were yellowish-white—the membrane is very rarely, if ever, of this color; they were isolated—the membrane forms at separate points, without coalescing, only in exceptional cases. These all recovered after three or four days, and their condition excited no apprehension for their safety so long as the diseased action went no further. The two other children had ash-colored membranous patches on the tonsils. In one, the membrane did not extend beyond the fauces, and although it fell off and formed again, the sickness was not very grave, lasting about ten days. In the other it penetrated into the larynx and proved fatal in three days. A physician's pride is in his cures, yet you may not be tempted to say, under parallel circumstances, that you had five cases of diptheria and lost but one of them. There were but two cases of diptheria. A physician reports that he has seen two hundred cases of diptheria, and has lost only three or four per cent. Another physician practicing in the same city, in the same epidemic, in the same class of society, and adopting the same general plan of treatment, has seen a hundred cases, and has lost thirty per cent. The latter states that he only counts the instances of membranous disease; the first considers his cases all diptheria, but does not say that membrane is his test. Would you not be compelled to hesitate before you admitted these two reports into the same generalization? Physicians who have had extraordinary success, should tell us in so many words, that all their cases, or if not all, how many, exhibited the membrane. You will not find this precision when you wish to ascertain, on a large scale, the law of mortality. Thus then is this problem complicated. But we must do the best we can with the material within our reach.

"From the 'Medical History, General and Particular, of Epidemic, Contagious, and Epizootic Diseases in Europe, from Remote Times to our own Days,' by Ozanam (vol. iii. p. 279), we learn that, so far as he could ascertain the facts, the mortality of thirty-nine epidemics of what is now regarded as diptheria, between 1559 and 1805, was as high as *eighty in the hundred* of those attacked. The Commission of the French Academy of Medicine (Martin Solon, and others) reported in 1833, that, in the French epidemics from 1771 to 1830, the deaths among those attacked by croup very often complicated with gangrenous angina, were as *one to four;* among those attacked by angina, membranous and gangrenous, simple or complicated, there was the same mortality, that is, *one in four*. ('Mem. Acad.,' vol. iii. p. 429.) Trosseau found, when pursuing his inquiries regarding cutaneous diptheria, that in some families and hamlets frightful havoc had been made by the throat disease. In one family *seven children had been attacked, and six died.* In one hamlet *ten in twelve* had died; in another *nineteen in twenty-one.* Dr. Thayer (*Berk's Medical Journal*) states that Dr. Beardsley had in Orange, Conn., among the pupils of a school, and in three families where the pupils boarded, *fourteen deaths in fifteen cases.* In the military school spoken of by Bretonneau, and referred to under the head of contagion, four deaths had occurred in as many cases, when the system was adopted of examining the throats of all the pupils daily; by this means sixty cases were seen at the commencement of the disease, and all were successfully treated; the mortality was thus, including a nurse that died, *one in thirteen.* Daviot states ('Memorial on Diptheria,' p. 363) that in the years 1841--2--3--4, he treated four hundred and sixty-one cases, and that he had forty deaths, a mortality a little better than *one in eleven.* Dr. Willard, in giving a history of the late Albany epidemic (*N. Y. State Med. Soc. Trans.*, 1859), feels authorized to reckon the cases in that city at two thousand, and the deaths as one hundred and eighty-eight, the ratio being *one to about ten and a half.* Dr. Kneeland (*Am. Med. Times*, Jan. 26, 1861) living in the central portion of this State, ascertained that among eighty persons attacked in his neighborhood ten died, or *one in eight.* Among the facts collected by Dr. Thayer are the following: six cases in Pittsfield, Mass., and three deaths, *one in two;* twenty-four cases reported by Dr. Bostwick, of Red Rock, N. Y., and five deaths, *one in five;* eighty-one by Dr. Meacham, of West Stockbridge, Mass., and eight deaths, *one in ten nearly;* one hundred and thirty-six by Dr. Wells, of Menomonec, Wis., and four deaths, *one in thirty-four;* eighty-five cases by Dr. Lawrence, of North Adams, Mass., and *no death;* 'forty or fifty' cases by Dr. Holmes, of South Adams, and *no death.* Dr. Jacobi, of this city, says (*Am. Med. Times,* Aug. 18, 1860), 'Of five hundred cases, we believe that we have lost not more than thirty,' about *one death in seventeen cases.* Dr. Watson, in a paper read before the New York Academy of Medicine, and published in the *Am. Med. Times,* states that of one hundred and forty-eight cases treated by himself, only two proved fatal, *one in seventy-four;* and in one hundred that he saw in the practice of other physicians of the city only four died, *one in twenty-five;* in all together about *one in forty-one.* Dr. Woodward, of Brandon, Vt, and his neighbor, Dr. O'Dys (*Am. Med. Times,* Dec. 5, 1860), treated *thirty cases each, without a single death,* and Dr. Woodward is careful to state that his cases were all true diptheria. In an adjoining town, where the disease occurred before it reached Brandon, he says *almost every case was fatal.* Statements that vary so widely as these do can not be usefully generalized. If they all relate to the same type of disease, they show the impossibility of applying a general average to particular epidemics. My own opportunities of seeing the affection force upon me a great mortality; but from what I have seen, and chiefly from what active practitioners here tell me, without any attempt at numerical computation, I should set down the deaths in New York among those having the *membranous* disease, including membranous sore throat in scarlet fever, as one in six or eight. But these conjectural estimates are worth very little. Indeed, accurate statistics will not avail much in informing you what to look for in any commencing or expected epidemic. I have already told you this, and I repeat it here because I wish you to be fully aware of the varying types of the disease. In reviewing what has just been said, you notice that, while in one place there is not a single death among sixty persons attacked, by what the writer assures us is true diptheria, in an adjoining town it destroyed nearly all whom it touched. This difference you may ascribe to different plans of treatment. I can not say it was not so in that instance. But you will hear it said that the disease is very grave in one place, and very mild in another; that the earlier cases in a school, hospital, or town are attended by a greater mortality than those which occur later in the epidemic. This is doubtless true, and it is true often because the type of the disease is different even in adjoining towns, and because its character changes as the epidemic advances. I am told by a leading physician of Massachusetts, that in a town within the range of his consultation practice, nearly every case is fatal, not by the direct effects of the membrane, but by prostration and collapse, without a sign of dyspnœa or cyanosis; while in another village five miles distant the disease has the characters of an open inflammation, from which the mortality is comparatively inconsiderable, and when death occurs it is almost always caused by laryngeal and tracheal obstruction through the extension of the membrane from the fauces. I have already quoted statements coming from two physicians of that same county, that they have treated one hundred and twenty-five cases without a single death, both saying that the disease was very mild (but neither of them saying that it was characterized by a membrane)."

COMPLICATIONS.

Among the numerous complications which are said to attend diptheria, authors mention *sthenic* or *high fever*, and *typhoid* or *low fever*. I must dissent from both of these propositions. The truth is this. The diathesis of the disease is essentially low or atonic, and hence can never have high or sthenic fever, neither as a part of the malady, nor as an accompaniment, nor as a complication. But as the disease is essentially febrile, and of low diathesis, and of continued type, it is always typhoid, so that this is an accompaniment and not a complication. The error of medical men consists in limiting their idea of fever to preternatural heat of the surface, and in mistaking a violent disturbance for strength of action. A fever, properly so called, consists of cold, hot, and sweating stages, these together constituting the paroxysm; and in all fevers excepting the ephemeral type, which lasts but one day, there is a succession of paroxysms; and this succession of paroxysms is the foundation for the nosological arrangement of fevers into intermittent, remittent, and continued types. But in many cases of low fevers the hot stage is so slight as to be scarcely appreciable, and the careless observer may regard the case as nonfebrile. When the heat of the surface is more considerable, yet slight and not uniform, the term "feverishness" is generally applied, or the "type" of the fever is said to be typhoid. And when the whole surface is preternaturally and decidedly hot, the fever is very apt to be confounded with entonic diathesis. But, as already explained, the disease is always febrile, and the fever, as well as the local inflammation, is always atonic and typhoid.

It is true that diptheria may supervene in the course of a simple typhoid fever; but more frequently it happens that the diptheritic exudation of the local inflammation does not appear until the constitutional symptoms which constitute the fever have been manifested for one or two weeks. The following cases, mentioned by Dr. Clark, in his lectures on this disease, are in point:

"Among the specimens of diptheria exhibited to you the present session, you will remember the tonsils, uvula, larynx, trachea, and fine divisions of the bronchial tubes of an adult lined by false membrane. The patient from whom the specimen was taken had been suffering from typhoid fever for two weeks at the New York Hospital, when he was attacked with symptoms of croup, and died in a few days, tracheotomy having been unsuccessfully performed. Several cases of a similar character were seen at the same hospital during the epidemic of typhus some years ago, in patients affected by that disease. It seems to have occurred in these cases after the completion of the second week of the fever. M. Louis ('Arch. Gen. de Med.,' tom. iv., 1824) has reported two cases of membranous exudation in the air-passages, and the usual symptoms of diptheria in patients having typhoid fever. One was a person twenty-three years old, who had been fourteen days in the hospital before the symptoms of the membranous disease began. The other was in a boy aged fifteen years. Dr. Greenhow ('On Diptheria' p. 76) reports that Dr. Heslop, of Birmingham, found in Nov., 1858, that of four cases of typhus fever occurring in one house, two of the patients had membranous exudation in the throat. In one of these it is stated that the patient, a girl aged seven years, had suffered nearly a fortnight before the appearance of the throat affection. In the other case the time of the occurrence of the latter is not mentioned. M. Louis' cases are described under the title, croup in adults; but as diptheria was prevailing in Paris at the same time, it is more reasonable to refer them to this class."

Albuminaria has recently attracted the attention of physicians as a complication of diptheria. Says Dr. Slade: "An element in the nature of diptheria is of recent discovery. We allude to the presence of albuminous urine in the disease. The first observation upon the relation of albuminaria to diptheria appears to be referable to a case reported by Mr. Wade, of Birmingham, to the Queen's College Medico-Chirurgical Society, in December, 1857, and afterward published in his 'Observations on Diptheria.' Shortly after this, during researches on this disease at Paris, M.M. Bouchut and Empis made a similar discovery. Albuminaria did not exist in every case examined, but it was seen in twelve cases out of fifteen. Both of these authors attach great importance to this renal complication, as affording an anatomical explanation of the cause of death, when this can not be attributed to either of the other modes, viz., death by asphyxia or general poisoning."

It seems to me that albuminaria can hardly be "an element in the *nature* of diptheria" unless it is invariably present; and if so it would be an essential part of the disease, and not, in any sense, an accident or complication; nor can I see the necessity for this "anatomical [pathological?] explanation" of the cause of death, when the patients do not die of "asphyxia or general poisoning," so long as exhausted vitality is a sufficient cause of death in all cases.

Other authors, however, have not found the complication of albuminous urine except in rare instances, and when present, they do not regard it as materially affecting the result. Prof. Clark, whose attention was some time since called to this subject, has only found albuminaria as an occasional accompaniment, and does not regard its occurrence as in any manner varying the prognosis.

Diarrhea, as we have seen, sometimes precedes and occasionally accompanies the diptheritic affection of the throat; and in some instances it attends the later stages of the malady, when it is regarded as an accident or complication. It is always a serious occurrence, as it indicates great exhaustion of the vital powers, and, consequently, danger.

Vomiting is regarded by some authors as a diagnostic symptom, and by others as an incidental occurrence, or complication. Though less dangerous than diarrhea, it is a troublesome and grave symptom.

Swelling of the glands of the neck, when extreme, is mentioned by Dr. Clark and some other authors as a complication, although, in a majority of cases, these glands are more or less swollen and inflamed. The chief difficulty arising from extreme enlargement of the glands, is the obstruction it occasions to respiration and deglutition.

"*Coma*," says Dr. Clark, "is an occasional termination;" but as it does not always terminate the disease, nor the life of the patient, it may more properly be regarded as an accident—a complication. It is, however, always an occurrence of dangerous import.

SEQUELÆ OF DIPTHERIA.

Almost all of the eruptive fevers, and more especially measles, scarlatina, and small-pox, are followed by many and often severe after-symptoms or secondary diseases, either the consequences of the disease, or of the treatment, or of both. And in this respect diptheria very much resembles them. Prominent among these sequelæ, authors mention various forms of paralysis, otalgia, amaurosis, ophthalmia, headache, etc.

"After apparent recovery from the immediate effects of the disease," says Dr. Slade, "in many cases, there still seems to be lurking in the system the morbid poison, whose special affinity is for the nervous system."

Such is the language of the author of the "Fiske Fund Prize Essay," on "Diptheria; its Nature and Treatment." And although the words are all in strict accordance with what is called the Medical Science of the nineteenth century, yet, judged by truly scientific principles, they are utterly nonsensical. A "morbid poison" implies the existence of a normal poison. But Nature teaches, and all the data of science, when correctly interpreted, affirm, that no poison is wholesome or normal, and hence no poison can need the qualification, morbid. *Poison is poison*, and that is all there is of it. Who would think of saying " black blackness," as though some kinds of blackness might be white, or of some shade between? It is true that there are *white blackberries*, and, in a certain stage of development, *blackberries* may be said to be *green* when they are *red;* but poisons do not undergo organic transformations, nor do they, under any circumstances of health or disease, change their relations to the living system.

But more absurd even than the notion of a "morbid poison," is the idea that it has a "special affinity for the nervous system." It seems to me, that if medical authors would look a little closer to the definitions of their technical words and phrases, they would not fill their books with such vague and meaningless, not to say false and ridiculous statements.

The only relation which a poison, be it "morbid" or otherwise, and the nervous system can hold to each other, is that of repugnance or antagonism, and this is exactly the opposite of affinity. In other words, instead of the poison having a special affinity for the nervous system, the whole living organism has a constitutional antipathy to the poison.

"The most frequent form of paralysis," says Dr. Slade, "has been that of the soft palate. The symptoms are, a nasal twang in the speech, incapacity for suction, and the regurgitation of fluids by the nostrils."

M. Trosseau states that, in consequence of the paralytic affection being more local than general—in other words, the palate and pharynx being more usually affected with paralysis than the system generally—he was for a long time under the impression that the loss of power was dependent upon the inflammation of the coats of the nerves supplying these parts, and an infiltration producing pressure on their motor muscles. A more extensive experience, however, of the general character of the paralysis which accompanies and follows diptheritic affections, caused him to change his views, and he now believes that loss of power and sensibility is the direct consequence of the peculiar diptheritic poison acting generally on the system, and strangely modifying the blood. M. Trosseau also states that, many children who have been subjected to the operation of tracheotomy fall victims to paralysis of the epiglottis and larynx.

Dr. Faure has more fully described the debility and paralysis which are so frequently supposed to be the sequelæ of diptheria, but which are, I fear much more frequently, the effects of the drugs which are administered for the cure of diptheria:

"Some time after an attack of diptheria, from which the patient has so completely recovered that no trace of false membrane is left behind, the skin grows more and more colorless without apparent cause, so that at length it assumes almost a livid pallor. Severe pains begin at the same time to be felt in the joints, the patient loses power over his limbs, and soon sinks into a state of indescribable weakness. At the same time, the disorders that appear in different functions show that the various organs which should minister to them are involved so far as they are dependent upon muscular power. In this respect, however, the phenomena are not constant, for sometimes it is one set of organs, and sometimes another which suffers most from this weakness. Very generally, in consequence of the want of muscular power, the patient becomes unable to sit upright, or does so with great difficulty, while the legs can not bear the weight of the body; all the movements grow uncertain, tottering, hesitating, and apparently purposeless. Very remarkable disorders show themselves also within the throat, for the velum is completely paralyzed, and hangs down like a flaccid, lifeless curtain, which interferes with speech and deglutition. All the muscles of the jaw, neck, and chest are partially paralyzed, in consequence of which mastication is rendered difficult, and the food can be neither easily moved about in the mouth nor readily swallowed. Vision is impaired, and squinting is not unusual. The sensibility of the skin is much diminished, in the limbs it is sometimes completely lost, though morbid sensations, such, for instance, as formication, are sometimes experienced. Œdema of the various parts often occurs, and occasionally parts here and there lose their vitality and become gangrenous. No general reaction occurs; fever is rare. The features grow duller and more and more expressionless, though a foolish smile sometimes crosses them, or now and then a ray of intelligence appears. Some patients have frequent fainting fits. As the condition goes on from bad to worse, the weakness becomes extreme, and death at length follows some fainting fit, or takes place when exhaustion has reached its uttermost; life, as it were, quietly, almost imperceptibly, passing away."

Dr. Greenhow remarks: " Under the most favorable circumstances, persons who have suffered from fully developed diptheria often remain feeble, ailing, and anæmic for many weeks; and the throat sometimes continues to present traces of the disease long afterward, or is very susceptible to the influence of cold or raw weather. Occasionally, many months elapse before perfect recovery; and I have known one instance in which the patient did not regain his strength for nearly a year. Besides the extreme anæmic which is so marked a result of diptheria, this disease is very apt to be followed by certain nervous affections of a peculiar kind. These consist of paralysis and anesthesia of particular muscles, tenderness and tingling of the skin, gastrodynia, impairment of vision, and deafness.

"Few persons recover without impaired voice or power of deglutition, arising from paralysis of the muscles of the throat; and sometimes, though rarely, there is complete aphonia, or absolute inability to swallow. The husky, nasal voice which follows diptheria is very striking, and closely analogous in character to that of persons suffering from syphilitic affection of the throat. It is remarkable that this affection, in common with the other nervous sequelæ not yet described, very often does not manifest itself until the patient is in other respects convalescent. The impaired power of deglutition consists sometimes of a difficulty in swallowing liquids, sometimes solids; but the former is the more common. Patients are sometimes able to eat a hearty meal without difficulty, but when they attempt to drink, a large portion of the liquid is regurgitated through the nostrils.

"The difficulty in swallowing liquids, and the nasal tone of the voice, are usually found in the same person; and although the voice is sometimes slightly affected without impaired power of deglutition, the latter is very rare without the former. Difficulty in swallowing solids, when the power of swallowing liquids is comparatively perfect, occurs but seldom.

"Paralysis of the muscles of the neck, producing inability to carry the head erect, is an occasional, but rare, sequel of the disease. Among a great many convalescents from diptheria that I have seen, not one has suffered from this affection."

The following case—interesting chiefly because of the time that elapsed between the affection of the throat and the development of the secondary disease—has been published by Dr. Gull, in the London *Lancet:*

"A boy, eleven years of age, had an affection of the throat from which he convalesced, and was sent into the country for change of air. About five weeks from the time of his being taken ill, it was noticed that he did not carry the head erect—it drooped to one side or the other. There was an occasional difficulty in deglutition, loss of voice, and attacks of dyspnœa, threatening asphyxia. In a day or two from the beginning of these symptoms, the breathing became entirely thoracic. The diaphragm was unmoved in inspiration and depressed in expiration, indicating a loss of power in the phrenic nerves. Deglutition was next to impossible. The child could utter no sound. There were fearful attacks of strangulation when the head was moved in particular positions, and even when the breathing was at the best, there were blueness of the lips and tracheal râles. The intelligence remained unaffected. The legs could be moved only feebly; the movement of the arms was not impaired; the muscles of the neck were wasted and flaccid; there was no swelling of the fauces; over the transverse processes of the cervical vertebræ, on the right side, there was tenderness, and the adjacent deep-seated absorbent glands were slightly enlarged; no febrile excitement. Pulse feeble, 90. A paroxysm of suffocation suddenly terminated the life of the patient."

A singular paralysis of the muscles of the neck, occurring after diptheria, is reported by Mr. Grundy, of Newick, in the case of his own son. The head rolled about by its own weight backward, forward, and sidewise, exciting fear of dislocation; and when it settled, the child was apparently unable to move it, and looked about him with a curiously slow turning of the eyeball.

"Paraplegia," says Dr. Greenhow, " is by no means an uncommon sequel of diptheria, and, though more rarely, paralysis of the arms. Sometimes the paralytic affection is of a *hemiplegic* character. The following case, which I had the opportunity of seeing with Dr. Morris, of Spalding, illustrates several of the points just mentioned, though the paralysis was less complete than in some other cases which I have seen : R. A., twenty-eight years of age, resides in a small but clean and wholesome house at Pinchbeck. His case was the worst that Dr. Morris had ever seen to recover. On Friday, January 28th, 1859, he felt a 'nasty taste' in the mouth. On the following day he complained of sore throat, and on examination by Dr. Morris it was found to be congested and inflamed. On the 30th, the tonsils, soft palate, and posterior fauces were covered with false membrane, and the case subsequently became one of malignant diptheria. March 20 : patient very pallid and anæmic; voice thick, snuffling, and nasal; there is a white filmy patch on either side of the arch of the palate, that on the right side being the largest; the uvula has nearly sloughed away, and he says that at the time of its occurrence the stench was so bad that he could scarcely bear it. On the right of the posterior fauces is a patch of opaque white false membrane, the size of a split pea; the rest of the posterior fauces is covered with a semi-transparent secretion. Skin sweaty ; pulse 72, feeble. Sight a little dim; complains of numbness in the belly, and in the legs, arms, and hands, but especially in the left arm and leg. Is unable to dress himself, from weakness of the arms; has lately felt pricking as of pins and needles in the fingers; is rather giddy when out of doors, and still has slight difficulty in swallowing. Three weeks since, his face was puffed in the morning, and there was slight edema of the feet and legs, particularly at night; urine pale colored, clear, and free from albumen."

It is to be regretted that, in the reports of these cases, no allusion whatever is made, except in rare cases, to the treatment. Says Dr. Bigelow, of Boston, in a late work ("Nature in Disease"): "The effect of remedies is so mixed up with the phenomena of disease, that the mind has difficulty in separating them."

I apprehend that the truth of this remark is quite as applicable to diptheria and its sequelæ, as to all other forms of disease. The ordinary drug-medication of diptheria is enough in many cases to paralyze not only the muscles of the neck, but those of the whole system, as is partially illustrated in the following case related by Greenhow, in which we have a hint of some of the remedial measures employed: " A woman, having been recently confined, contracted diptheria from a patient in a neighboring bed. *Alum in sufflations* and applications of *hydrochloric acid* were resorted to, with the effect of removing all diptheritic exudation. On the tenth day she spoke with a nasal voice, and deglutition was very difficult, and accompanied with nasal regurgitation. A notable proportion of albumen was also found in the urine. The paralytic affection of the pharynx kept increasing, so that by the twenty-fifth or thirtieth day the woman could no longer swallow, and was like to have died while trying to take some solids. About the fortieth day some improvement in this respect took place, but some numbness of the hands and feet was observed, as well as defective pronunciation from imperfect movement of the tongue. By the fiftieth day, progression had become uncertain, and general nervous symptoms, chiefly consisting in delirium and convulsions, set in. The worst apprehensions were now entertained; but musk having been administered, some improvement took place. So considerable, however, was the paralysis, that the patient could not raise herself without the assistance of two nurses. The bladder was also affected during two or three days, but not the rectum. With this paralytic condition anæsthesia coexisted, the patient remaining absolutely insensible to pricking with needles. On the hundred and fiftieth day the symptoms were so much ameliorated under the use of the syrup of the *sulphate of strychnia*, that the patient could get in and out of bed easily, could knit a little, and was able to distinguish between wool and cotton by the touch. No disturbance of visual power took place, although during six weeks enormous quantities of albumen were found in the urine."

The medication in the above case, though far from being as potent as is frequently prescribed, is amply sufficient, in my judgment, to account for all the complications and sequelæ which afflicted the unfortunate woman, and for the protracted convalescence. All of the caustic and burning, pungent, local applications, including nitrate of silver, chlorate of potassa, alcohol, etc., are of paralyzing tendency, and any variety or *quantity*, if I may be allowed the expression, of " general nervous symptoms," may be justly attributed to their employment; and when the effects of these remedies become " mixed up" with the phenomena of disease, I know of no way in which the physician can separate them.

Because the patient, after lingering one hundred and forty-nine days, in virtue of an enduring constitution, improved while taking the deadly dogbane, or because her symptoms became ameliorated while "*under the use*" of this drug, it by no means follows that the strychnine contributed to the amelioration of the symptoms. On the contrary, any one who can reason from the physiological instead of the pathological stand-point —who can interpret the effects of remedies and the phenomena of disease by the unerring standard of the laws of Nature as manifested in and through the vital organism, instead of by the false and absurd dogmas of medical schools, as taught in their books on materia medica—will know absolutely that all such patients, when they improve or recover, do so in spite of the medicine. No person whose brain is not prepossessed and prejudiced by the false theories of the day which pass current in the world as medical science, can read the "*modus operandi*" of strychnine, as stated in any of the standard works on materia medica, and not come to the conclusion that its effects are in every case, and stage, and condition of diptheria, as well as in all of its complications, incidents, accidents, accompaniments, concomitants, or sequelæ (and the same is equally true of all other diseases), to prolong the patient's sufferings, lessen his chance of final recovery, and render recovery less complete.

"*Impaired vision*," says Dr. Greenhow, "is another common sequel of diptheria, which, like those already described, only comes on subsequently to recovery from the primary local disorder. The patient is usually able to see distant objects with sufficient distinctness, but is unable to see things close at hand. Indeed, several of the most striking cases that have come under my notice were those of children who appeared to be quite well until, on returning to their studies, it was found that they could not see to read. The defective vision comes on gradually; first of all, the patient is unable to read small print, and can only read large print when held at a distance from the eye, a power which is also lost at a later period. The restoration of sight is equally gradual."

As I have never noticed any serious difficulty of vision as a sequel of diptheria, in cases where the patient has been treated hygienically, and as quinine and other similar drugs are well known, when given in large doses, to affect the vision very seriously, I am apprehensive that this "sequel" of impaired vision has some definite relation to the medication. Many practitioners recommend the free use of quinine and other "supporting" agents throughout the whole course of the disease. Indeed, the plan of treating typhoid and other low fevers with brandy and quinine, from the commencement of the disease to the end of the convalescence, on the senseless vagary of "keeping the patient up" while the disease runs its course, or on the equally chimerical fantasy of "carrying the patient through the disease," has recently been revived by Dr. Todd, of England, and some other practitioners, so that we may soon look for impaired vision—and also for *deafness*, which is a very common effect of quinine—among the very common sequelæ of an extensive range of febrile maladies.

As an illustration of some of the *nervous sequelæ*, the case of Dr. Moyce, of Rotherfield, Eng., is related: "On Nov. 8, 1858, he felt a sensation of pricking, which soon became burning, in the right tonsil. In the night there was much pain, with a sense of swelling. The next morning there was on the right tonsil a patch of exudation about the size of a farthing, which gradually extended forward almost to the teeth; the left side was very slightly affected. There was much external swelling. After four or five days the exudation began to clear away, and then pain and difficulty in swallowing, amounting to agony, supervened. In the course of three or four weeks he got about, and attended to his practice for a fortnight. During the latter half of December the tone of his voice became altered, and he began to have regurgitation of solid food, which would accumulate in the posterior nares until it caused spasmodic cough. He was able to swallow fluids, if taken very slowly. He now lost the use of his tongue, could not move it in eating, and his speech became unintelligible; he also began to see double, and indistinctly, but could see with spectacles. Next followed tingling and tenderness of the palmar surface of the hands and fingers, accompanied with a peculiar hardness and roughness of the integuments. Presently the soles of the feet and toes were similarly affected, and then there was loss of power in the limbs, especially the legs. The arms were so weak that he was unable to feed himself. The symptoms remained unabated for eight or nine weeks, and then gradually diminished in the same order in which they had begun. Even now, after a lapse of two years and a half, he is not strong, and can neither walk nor swallow so well as before his illness."

It is deeply to be regretted that in so extraordinary a case affecting a medical man, not a word nor a hint should be given in relation to the treatment. If the Doctor dosed himself with calomel, quinine, stimulants, and tonics, alteratives and antiseptics, continually, the "nervous sequelæ" and the prolonged convalescence may be easily accounted for.

The following case, related by Dr. Gull, is suggestive of *mercurialization*, though not a word is said as to the treatment:

"A boy, of rather delicate temperament, when recovering from diptheria, was suddenly seized with intense *neuralgia* in the left leg, which passed off after a day. It appeared to be connected with the femoral vein, which was rather hard and very painful to the touch. After two days he became very restless, and, in a few hours, completely *hemiplegic* on the right side, including the face, and speechless. The action of the heart was most tumultuous, and the sounds muffled. The child rallied under the free use of wine and ammonia; but the hemiplegia remained for many months, after which there was slow improvement."

This "rallying" under the use of stimulants is one of the great delusions of the medical profession, and of the non-professional people. The effect or disturbance which is called "rallying," or "reaction," is the resistance of the living system to the poisons. *It is the drug fever;* nor is it any the less injurious to the patient because it is termed *stimulation,* and is caused by a drug which is termed *medicine,* than it is when it is called *disease* or *fever,* and is caused by a drug which is called a "morbid poison."

Says Dr. Greenhow: "The majority of cases which are protracted until the development of the nervous sequelæ, recover, but death occasionally takes place even at a remote period. Dr. Moyce mentions the death of a boy, aged eleven or twelve years, from exhaustion during the paralytic stage, two months after he had been quite free from throat affection."

In view of the ordinary treatment of diptheria, I think the above remarks should be understood as meaning, if the patient can survive both the disease and the remedies, he may recover sooner or later, although he may long suffer from the chronic disease induced by the medication.

And still more to the point says the same author: "The nervous sequelæ of diptheria are not always in proportion to the severity of the previous illness, and do not occur exclusively after the severest cases, but sometimes follow comparatively mild attacks. Their duration is uncertain, varying from two to three or four months, but the slighter affections may perhaps sometimes pass off in a shorter period than two months, and, in all probability, severe cases are occasionally prolonged beyond the fourth month."

If the sequelæ were legitimately the consequences of the diptheria, or of the causes of diptheria, it would logically and necessarily follow that the more severe the disease the greater the liability to sequelæ, and the more severe the sequelæ. But if the sequelæ are chiefly attributable to the remedies employed for the cure of the diptheria, then we may properly expect just what our authors inform us is the fact, that mild cases of the disease may be attended with dangerous complications, or followed by severe sequelæ, and *vice versa*.

Bronchitis and *Pneumonia* are named by some authors as complications, and by other authors as sequelæ of diptheria. But I think the bronchial affection is merely the extension of the diptheritic exudation to the bronchial tubes; and that the pneumonia is the same with a more considerable degree of congestion in the lungs—a condition which may occur in the dying struggle. And these views are corroborated by all the circumstances of the cases adduced to prove the existence of these affections. Mr. Thompson reports the following case, in the *British Medical Journal* for June 5, 1858: "A gentleman, aged forty-six, died from this condition of the lungs. His throat was first affected. After a few days the breathing became impeded, with all the ordinary symptoms of capillary bronchitis in the first stage, the throat continuing to improve. He gradually sank, constantly expectorating casts of the small tubes, precisely similar to the deposits in the trachea."

Drs. Greenhow, Bristowe, and others, state that they have only found the occurrence of pneumonia as a complication of diptheria has only come under their observation in *post-mortem* examinations. Mr. Rush, of Southampton, mentions two cases in which fatal pneumonia supervened after the exudation had disappeared from the throat, and the patients were supposed to be doing well.

The occurrence of fatal secondary diseases, long after convalescence in relation to the primary disease has been established, is always, to my mind, suggestive of *drug-disease*. And does not Professor Paine, in his "great work" ("Institutes of Medicine"), declare, as the basis and rationale of the whole drug system of medical practice, " we do but cure one disease by producing another."

MORBID ANATOMY OF DIPTHERIA.

Post-mortem examinations can never reveal the *essential nature* nor the *causes* of any disease; they can only exhibit the *effects* of disease—the morbid conditions which occur in its progress, and the structural derangements which take place after death. But these effects and derangements may be the results of the disease itself, or of the medication, or of both. And if it is very difficult to discriminate between the phenomena of disease and the effects of remedies in the living subject, it is still more difficult to determine, in the cadaver, what appearances are due to the original disease, or to its causes, or to the medicines, which are themselves morbific agents, and must of necessity induce disease. Hundreds of *post-mortem* examinations have been made after deaths of pneumonia—inflammation of the lungs; and when mercurial and antimonial remedies had been prescribed, there have been found as complications and sequelæ, morbid conditions of the stomach and bowels to which the terms *gastritis* and *enteritis* are applicable. Whence this inflammation of the stomach and bowels? Dr. Ames, of Montgomery, Alabama, in an article published in the New Orleans *Medical and Surgical Journal*, a few years ago, states that these complications are found only in cases which have been treated with bleeding, calomel, tartar emetic, and other powerful drugs—never in cases treated with what are called simple remedies or mild medicines.

What do these facts prove? What can they prove, except the admitted fact that all drug-medicines are poisons, that all poisons induce diseases, and that when poisons are administered as remedies to cure diseases, we must of necessity find the effects of remedies and the phenomena of disease so "mixed up"—to quote again the language of Dr. Bigelow—as to render it exceedingly difficult to distinguish the one from the other.

In estimating the value of pathological anatomy, we must ever keep in mind that the dead structures can only disclose the effects of morbific agents and processes; they can never explain the remedial actions—the vital struggle—which constitutes the very essence of disease. My work would be incomplete without a chapter on this subject; and as Dr. Greenhow has presented in his late work all of the facts pertaining to the morbid anatomy of diptheria which are known to the profession, or which can be of use in determining either the nature, the causes, or the proper treatment of the disease, with illustrative cases, I copy his entire article, premising that a careful examination of all evidences can hardly fail to convince the candid reader, especially if he is familiar with the effects of medicines as explained in the works on materia medica and therapeutics, that many of the morbid appearances described are quite as likely to be the effects of drug-poisons as of diptheria or its causes.

"Diptheria is essentially an inflammation of the fauces, which sometimes only causes disordered secretion from the mucous membrane; at others, produces ulceration, and even gangrene; but, more frequently, an exudation which, coagulating on the surface, forms the false membrane from which the disease obtains its name. The exudation varies in consistency from a pultaceous or almost liquid exudation to a firm, consistent, and more or less elastic membrane. In the latter case, its outer surface is often uneven, usually less dense than the deeper portion, and sometimes flocculent or fissured. It varies from a quarter of a line to a line or more, and, in one instance I have seen, was nearly two lines in thickness. The elastic form of false membrane is not unlike the exudation poured out from an inflamed serous membrane. Sometimes the exudation is not membranous, but dry and granular.

"Low forms of cryptogamic plants are occasionally found on the exudation, a circumstance which gave rise to the belief that the disease is of parasitic origin. This opinion is disproved by the facts that, on the one hand, the supposed parasite is not invariably present in diptheria; and, on the other, that it is frequently found on unhealthy mucous surfaces which are not of a diptheritic nature. Examined under the microscope, the exudation is found to consist of coagulated fibrine and epithelium, the latter being usually more abundant in the outer portion, or layer of membrane; while the deeper portion is more purely fibrinous. But in this respect there are numerous variations. Exudation cells are often intermixed with the fibrillated texture. The exudation is sometimes already undergoing decomposition, or other change, before it leaves the throat, and is at others more or less stained with blood. At first only opaque, the exudation soon becomes white or ash-colored; if thick and adherent, brownish or buff-colored; and if stained by slight hemorrhage, blackish. The exudation is sometimes very loosely, at others very firmly, adherent to the subjacent surface; and occasionally, especially when of the friable, granular variety, is merely superimposed upon the natural surface.

"The mucous membrane underneath the exudation, or from which the exudation has recently exfoliated, is often intact, and generally much congested and swollen; sometimes it is white, opaque, or unnaturally pale; at others it looks raw, the epithelium having been shed with the false membrane. It often presents an excoriated and roughened appearance; is sometimes ulcerated, and, more rarely, gangrenous. When false membrane, still adherent to the mucous surface, is lifted up, it is often seen to be attached to the subjacent surface by numerous small thready adhesions, as though processes of exudation passed into the mucous follicles; and, on removing it, the mucous membrane is more or less abundantly dotted with bloody points.

"The submucous tissue is often edematous, infiltrated with blood, and sometimes the seat of interstitial exudation. The tonsils are usually swollen, and, on being cut into, are often infiltrated with blood, so as to impart to them an ecchymosed appearance; sometimes their tissue is softened; and in two instances I have found the center of a tonsil in a state bordering on gangrene. There is generally more or less of inflammatory effusion into the structure of the tonsils; and in one instance, on the tonsil being laid open, there was an oozing from it of a creamy fluid resembling pus. In some instances, the esophagus and the muscular and other tissues around the fauces are congested or infiltrated with blood; the parotid and submaxillary regions are much swollen, and the integuments studded with livid purpurous spots. In a case mentioned to me by Mr. Jauncy, of Birmingham, an abscess was found between the pharynx and vertebræ. The case was that of a child, aged six years, which died after an illness of nine or ten days, croupy symptoms having set in three days previous to death:

" 'The lungs were emphysematous in front, collapsed in patches posteriorly. A portion of false membrane was found at the bifurcation of the trachea, which was elsewhere free from exudation, but reddened. The larynx, epiglottis, pharynx, tonsils, and uvula were covered with lymph. An abscess about the size of a walnut was found between the pharynx and vertebræ. Liver, kidneys, and spleen healthy. The kidneys were examined microscopically.'

"When the disease extends to the larynx and trachea, the false membrane generally becomes thinner and less consistent as it descends in the tube, until it disappears gradually in the form either of a very thin pellicle, or of a creamy fluid. The mucous membrane of the affected portion of the larynx and trachea is generally more or less congested, and often thickened, so as to diminish the caliber of the passage, even after the false membrane has been removed, or has come away. The subjacent membrane is here, for the most part, intact; but sometimes, being denuded of its epithelium, exhibits, on the removal of the exudation, a red excoriated appearance, somewhat like the raw surface produced by a blister. It also, under the same circumstances, presents small bloody points similar to those observed on the mucous membrane of the pharynx. The epiglottis, besides being covered above or below, or on both sides, with exudation, is likewise often swollen so as to contract the entrance to the windpipe. The bronchial tubes are sometimes lined with false membrane down to the third or fourth bifurcations, and even lower; and the lungs, sometimes partly emphysematous, are also liable to be affected with pneumonia, which is most commonly of the lobular form. In the latter case, the little bits of splenified lung are sometimes surrounded by crepitating and comparatively healthy lung, sometimes by portions of emphysematous lung.

The kidneys have sometimes been found quite healthy after death from diptheria; in other cases they have been congested, and, on being sliced, have exhibited under the microscope transparent fibrinous casts of the tubes. The urine, in such cases, is generally albuminous, and also presents under the microscope fibrinous casts of the tubes, which occasionally contain blood corpuscles, or granules of hematine, or a few altered epithelial cells.

"In a case briefly referred to by Dr. Gull, in his communication to the medical officer of the Privy Council, the membranes of the brain and cord were in a state of suppurative inflammation, the subarachnoid space being full of soft, purulent lymph; and the same physician, although he gives no *post-mortem* facts in support of the opinion, suggests that the original seat of the disease being near the cervical portion of the spinal cord, the paralytic symptoms so common in a late stage of diptheria may arise from the disease having extended by continuity, from the fauces to the upper part of the cord. At present, this opinion can only be received as suggesting a careful examination of the cord in future *post-mortem* examinations; for thus only can it be determined whether the paralytic affection has a constitutional origin, or arises from the supposed local disease.

"In a case related by Dr. Bristowe, and exhibited by him at the Pathological Society, the muscular tissue of the heart was colored with extravasated blood. And in a more recent case, treated by the same physician in St. Thomas's Hospital, in which I had the opportunity of examining the organs after death, the heart was studded with petechial spots on its outer surface.

"The following cases are adduced in illustration of some of the points mentioned in the preceding account of the morbid anatomy of diptheria. The first has been selected because it well shows the tendency of the disease to become engrafted, so to speak, on other disorders, especially the eruptive fevers; the others, mainly on account of the detailed description of the microscopical appearances noted by such competent observers as Mr. Simon and Dr. Bristowe.

"S. Beard, aged four years, was admitted a patient of the Western General Dispensary, under the care of my colleague, Dr. Sanderson, on June 29, 1859. She had been taken ill on the previous day with the premonitory symptoms of measles, and was visited by the house surgeon, Mr. Plaskitt. It was not until the 4th of July that she complained of her throat; and she first came under the observation of Dr. Sanderson on the 6th of that month. The skin was then of a not unnatural warmth; the countenance was pale, and its expression rather distressed. The child was somewhat drowsy, and difficult to rouse; there was a slight discharge from the nostrils, which were lined with coagulated blood arising from an epistaxis on the previous day. Respiration natural in frequency; pulse 120; the mucous membrane covering the tonsils was of a deep red color, but less bright than is usual in ordinary tonsillitis. The anterior surface of the uvula was bare, but the posterior surface and sides were covered with a soft concretion, capable of being detached, and evidently of slight consistence. All the parts were smeared with a tenacious mucus, which was constantly being discharged from the mouth; and flakes of concretion, which had been excreted during the preceding night, were exhibited by the mother. There was very little external swelling or tenderness about the neck, and the breathing was not at all croupy, although said to have been so. Urine intensely albuminous.

"*July* 7.—A tubular cast, of soft consistence, distinctly marked by the laryngeal rings, was discharged during the night.

"*July* 8.—Much worse; feet and hands warm; belly hot. Pulse 160, feeble, and very difficult to count; respirations about 30. Prolonged, somewhat musical expiration sound, varying in tone from minute to minute; inspiration sound, short, less noisy, and not musical. Countenance pale, but not livid. Voice resembled a shrill whisper heard through a long tube. The cough, which occurred occasionally, was very short, and precisely similar in tone to the voice. A few small shreds of concretion were still attached to the uvula and velum; but none elsewhere. There were excoriations at the corners of the mouth, not covered with concretion. Mucous surface of a deep-crimson hue.

"*Vespere.*—Respiration increased in frequency to 40 in the minute; countenance more indicative of distress. She died at seven A.M., on the 9th.

"*Post-mortem Examination* (made June 10, twenty-seven hours after death).—Slight mottling on the arms, probably the remains of the eruption of measles. The upper surface of the tongue was healthy as far backward as the base of the epiglottis, excepting that there was a small patch of exudation, not much larger than a grain of wheat, adherent to one of the large papillæ. The subjacent surface was healthy; both tonsils, especially the right, were vascular, and presented a pitted, roughened appearance. The mucous membrane covering the margin of the epiglottis, epiglottidean folds, and arytenoid cartilages, was white and opaque. The anterior portion and edges of the upper surface of the epiglottis were of a brownish-white color. The mucous membrane of a cavity behind the left tonsil and between it and the posterior pillar of the fauces contained a creamy-looking exudation. The corresponding hollow on the right side was free from exudation. The substance of the tonsils, particularly of the right, was decidedly softened. On being incised, they exhibited patches of extravasation and of pigmentary discoloration; but in other respects the section presented a natural aspect. The mucous membrane of the larynx and trachea was unnaturally white and opaque, as though covered with exudation; but nothing could be stripped off it. This condition of the membrane became less and less obvious in a downward direction. Here and there were seen punctuated patches of redness, which sometimes followed the intervals between the rings of the trachea. Several loose fragments of exudation, some of which, although readily detached, were still adherent to the natural surface, were found in the upper part of the trachea. The subjacent mucous membrane was unbroken, and closely resembled the surrounding mucous surface.

" The apex and upper portion of the left lung, as far as a line extending upward and backward from the notch, was emphysematous, and along the free margin were emphysematous lobules, surrounded by portions of splenified lung. The lingua and margin of notch were completely splenified. The secondary division of the bronchus leading to the apex of the left lung contained cylindrical casts, of about the consistence of boiled macaroni, at their proximate extremity, but diminishing in consistency until they disappeared in the third or fourth division of the bronchus, in the form of creamy-looking fluid. The division of the bronchus leading to the lower lobe contained no casts, excepting in one of the tertiary divisions leading toward its posterior aspect. It was not ascertained whether or not this portion of exudation was continuous with that in the bronchus leading to the apex.

The mucous membrane was for the most part remarkably pale, but otherwise healthy. There was bronchitis in a few of the smaller tubes, as shown by the frothy secretion which they contained, and by slight vascularity. The parenchyma was firmly splenified throughout the lower lobe, with here and there scattered portions of emphysematous lung.

"The two upper lobes of the right lung were emphysematous; the lower lobe was also emphysematous at the upper portion, and partially so below. The bronchus leading to the apex contained here and there adherent, but also partly detached, patches or fragments of soft exudation, which ceased rather abruptly in the third bifurcation, and less decidedly terminated in creamy fluid than those on the left side. A considerable-sized tube leading toward the base of the upper lobe was choked with a cylindrical mass of semi-diffluent white and opaque secretion, which, under the microscope, exhibited cells without fibrinated matrix. The bronchial branches leading to the middle and lower lobes were free from exudation. The mucous membrane of the tubes in the upper lobe, like that on the left side, was perfectly white. That of the tubes leading to the middle and lower lobes on the right side markedly injected.

"The following case, communicated to the Pathological Society by Mr. Simon, is quoted from the Transactions of that Society for last year:

"'A. H., æt. thirteen, had been suffering from diptheria for nineteen days before his death, and during the last eleven had been under treatment in St. Thomas's Hospital. On the eighth day of the disease a large mass of thick, dense, very fibrinous false membrane detached itself from the fauces, leaving the surface of the tonsils and soft palate raw (like that of skin from which the cuticle has been removed after blistering), but not ulcerated or sloughing. On part of this surface, a second thinner false membrane soon formed, and subsequently came away in shreds. There was irritating discharge from the nose, and during the last days of life some of the patient's drink escaped this way. Early in the disease there had been swelling below the jaw, but this had subsided many days before death. On the seventeenth day of the disease superficial ulceration began at the left tonsil, and on the eighteenth day had extended to the size of a shilling. On each of the last eleven days of life the urine was examined; it always gave abundant precipitate with nitric acid, and latterly also with heat; but in the earlier days it precipitated imperfectly with heat, and largely with acetic acid. Microscopically it showed fibrinous tubule-casts, containing traces of hemorrhage, but scarcely any renal epithelium. Throughout the progress of the disease the patient was pale, feeble, and disposed to be chilly, so that wine and much external warmth had from the first been necessary. The tongue was always moist. No eruption appeared upon the skin. There was no delirium or stupor, and neither cough nor any sign of laryngeal obstruction was observed. The respiration was natural till within a few hours of death, when it became short and hurried.

" 'The following were the *post-mortem* appearances: With the exception of an occasional very delicate film, there was no false membrane about the fauces. In the situation of the left tonsil was a sloughy ulcer, somewhat larger than a shilling. The posterior surface of the soft palate was congested, and there adhered to its somewhat swollen mucous membrane small patches of false membrane. In the recess of mucous membrane beside the epiglottis was an irregular depression, evidently the remains of an almost cicatrized ulcer. About an inch below the aperture of the glottis, the pharynx presented on its right side a small circular ulcer, about two lines in diameter, with somewhat raised margins, and on the left side another similar ulcer, about the size of a pin's head. In other respects the pharynx and esophagus were healthy. On washing out the nares a strip of false membrane an inch in length was removed. The mucous membrane covering the septum showed patches of congestion, was thickened, and had shreds of false membrane adherent to it.

"'Both lungs, except in their upper and anterior parts, were greatly congested with blood, and less crepitant than is natural, especially the lower lobes, whose posterior parts were in many places nearly or quite without air; and the most solidified portions broke down on firm pressure with the finger. At one section the exuding fluid was obviously purulent, and microscopical examination showed pus extensively in other parts of the hepatized structure. The bronchial mucous membrane was a little injected; the tubes contained thin frothy fluid tinged with blood, or more tenacious reddened mucus.

"'The kidneys were large, and intensely congested. Sections of the cortex, microscopically examined, showed frequently the presence of large, transparent, colorless rods of apparently fibrinous material, soluble in acetic acid and liquor pottassæ. These rods were sometimes floating free, sometimes partly or wholly held within urinary tubules, of which evidently they were casts. They were generally structureless, but (no doubt from the manner of their formation) had a disposition to transverse fracture, and sometimes presented lines curving almost concentrically across them, or had this direction given to little clusters of granular matter, probably altered epithelium, which they occasionally contained. Apart from the presence of these casts, the tubular structure of the kidney was not very obviously diseased; but, after prolonged and careful observation, it could confidently be said, that, at least in many parts, the cell-growth within it was redundant, so that the tubules were more opaque than natural, and had their interior canal encroached upon, or even quite occluded by an increased amount of epithelium. The Malpighian tufts within their capsules showed a little indistinctly.

"'The venous system was everywhere remarkably full of blood; the liver was greatly congested; the heart was healthy, with a firm coagulum in each of its four cavities.'

"The next case, also taken from the *Transactions of the Pathological Society*, is from a communication by Dr. Bristowe:

"'T. N., æt. ten, the son of a farm-laborer, was admitted into St. Thomas's Hospital, under Mr. Solly's care, on the 12th of November, 1858, with contraction of the left wrist and elbow-joints, after a burn. On the 18th he was operated upon, and continued under mechanical treatment up to the commencement of the malady of which he died. He appeared perfectly well on the 20th of March, 1859, but on that day partook of some gin and other improper articles of diet. The following morning he had a slight attack of shivering, and seemed otherwise a little indisposed. On the 22d he complained of slight soreness of the throat. This increased, and on the 24th the following notes were taken by the surgical register:

"'Throat much swollen externally, particularly on the right side. On looking into it, the right tonsil is seen filling up the fauces, and has upon it a pultaceous material. Pulse small and weak, 130; tongue furred; skin cool.'

"'On the 25th he was placed under my care. He has slept a little in the night, and is said to be now rather better than he has been. He is extremely feeble, however, not at all feverish, and perfectly rational. The skin is cool, and gives no indication of rash. Pulse small, weak, slightly irregular, and about 100. There is great tumefaction, hardness, and tenderness in the upper part of the throat, chiefly in the parotid and submaxillary regions, and more on the right side than on the left. The anterior half of the tongue is clean, and its papillæ are healthy; the posterior half is somewhat furred. The right tonsil is much swollen, and covered by a thick wash-leather-like false membrane, which is prolonged from it on to the pillars of the fauces, over the right half of the soft palate, and to the edges of the posterior teeth. The nose bled this morning, and a little thin sanious fluid has continued to ooze from it. Has no pain anywhere except in the throat; experiences pain and difficulty in swallowing, but can manage to take fluids. No cough or difficulty of breathing. Bowels opened yesterday.

"'*March* 26, two P.M.—Slept pretty well, but is much worse than he was. Skin cold, without trace of rash. Pulse quite imperceptible. Throat more swollen, hard, painful on pressure, and studded on the right side with small congested points. Tongue dryish, but not much furred. The breath has a faint, gangrenous odor. There is no appreciable change in the condition of the interior of the throat. Is quite sensible, but very restless. No cough, or embarrassment in breathing. He continued to sink, and died at half-past five, P.M., remaining sensible to the last.'

" The following were the *post-mortem* appearances:

" 'The front and sides of the throat were thick and brawny; and the parotid and submaxillary regions were much swollen and hardened, especially on the right side, where also the integuments were studded with congested and livid spots. On cutting into the neck, its muscular and cellular tissues, from the integuments to the vertebræ, and from the ears and root of the tongue to the upper opening of the thorax, were found indurated and brawny, and so infiltrated with blood as to be everywhere almost black. There were no circumscribed fluid or clotted collections, but the blood was uniformly diffused throughout the tissues. There was no appearance of pus, and no visible indication of inflammatory deposit.

" 'The soft palate and uvula, the tonsils and pillars of the fauces, the esophagus and larynx, were all intensely and deeply congested, tumid, brawny, and covered in many places by toughish, adherent, ashy, false membrane, or by pultaceous puriform exudation. The soft palate was quite half an inch thick, infiltrated with blood, and studded with shreds of false membrane. The tonsils were swelled, but at the same time presented deep fissures and excavations, and were covered pretty completely by grayish-yellow false membrane. This was in parts thick, tough, and pretty firmly adherent; but over the convexity of the tonsils became changed into a soft, pultaceous deposit, which seemed partly pus and partly superficial slough. On incising the left tonsil it was found softened, deeply congested, partly infiltrated with blood, and studded with distinct pus-holding cavities; and the surfaces of the fissures passing into it from the surface were soft, greenish, and slightly gangrenous. The right tonsil was generally in the same condition as the left, but presented several deep, distinctly gangrenous, fetid excavations. The mucous surface at the base of the tongue and back of the pharynx was congested, and presented here and there shreds of adherent membrane. The mucous investment of the epiglottis, and indeed that of the whole larynx, were thickened, indurated, and deeply congested. The epiglottis was covered pretty extensively by a toughish adherent membrane, about half a line thick; and a similar formation, in less abundance, was studded over the rest of the laryngeal surface, and accumulated along the vocal cords. The trachea was congested, but otherwise healthy; the esophagus also was healthy; but the tissues immediately surrounding them, like those of the rest of the neck, were infiltrated with blood. Several portions of the hard palate, and septum nasi, were removed, and their mucous covering was found congested, and lined by adherent false membrane.

"'Pericardium healthy. Heart small, firmly contracted, and nearly empty, its auricle and right ventricle containing a little fibrinous clot only. The valves were healthy. The muscular tissue was generally pale; but almost all the musculi papillares and carneæ columnæ of the left ventricle, and the walls of the apical half in nearly their whole area, and to a depth varying irregularly from a quarter of an inch downward, were almost black from sanguineous infiltration. The same condition was observed in the right ventricle, but to a less extent, the papillary muscles and the parietes being studded irregularly and thickly with black, blood-infiltrated patches of various sizes; some so thick as to reach the external surface of the organ, and some dotted with white spots and patches, which looked at first sight like suppurating points.

"'Pleuræ healthy. Lungs crepitant throughout, and not materially congested. They presented, however, on their external surface, a few dark-red, almost black spots, about a quarter of an inch in diameter, which were found to correspond to small subjacent patches of solid, dark-colored, granular lung tissue. The bronchial tubes contained much secretion.

"'Peritonium healthy. Liver of usual size, generally of normal color and consistence; its surface and substance, however, were thinly studded with petechial spots. Spleen of usual size, pale, and of moderate consistence. There was a little effusion of blood in the sub-mucous and cellular tissues around the pancreas and supra-renal capsules; and the latter organs presented patches of extravasated blood in the interior, though apparently in other respects healthy. The cellular tissue of the mesentery was studded pretty thickly with small, and not very intensely-colored patches of congestion and extravasation. The stomach and intestines were healthy, but the ilium contained two lumbrici. The kidneys were of the usual size, pale, and apparently perfectly healthy. Aorta and vena cava healthy.

"'The false membrane about the fauces and neighboring parts was made up chiefly of a net-work of fibrillated lymph. The fibrillæ were very irregular in outline and dimensions, but generally comparatively thick; and they coalesced with one another in all directions, so as to leave irregular spaces between them, which were small, and often not larger in diameter than the fibrillæ themselves. When seen in thickness, the tissue above described presented a pebbly character, like that afforded by an accumulation of nuclei; but the fallacious nature of this appearance was recognized on looking at the thin edge of a section; or by adding acetic acid, which rendered the whole transparent, at the same time expanding it, and bringing into view an exceedingly delicate and irregular net-work of well and sharply-defined, occasionally bulging, fibers, which appeared to be, so to speak, the skeleton of the original net-work. In some places the false membrane consisted of an apparently uniform layer, composed of an extremely fine and indistinctly fibrillated tissue, studded with molecular matter, and presenting something of a ground-glass character. Imperfect epithelium was entangled here and there in the substance of the membrane, but was most abundant on the superficial surface.

"'The pus-like fluid in the tonsils consisted of well-marked pus-cells characteristically affected by acetic acid. Some of the muscular tissue from the small muscles of the larynx and from those of the neck was examined, and found to be striated and healthy-looking; but the spaces between the fibers were loaded with blood-corpuscles. The cellular tissue in front of the epiglottis presented a net-work of fibrillated tissue like that constituting the false membrane itself; but the meshes were larger and more distinct. The muscular tissue of the heart was found to be generally in an early stage of fatty degeneration, the transverse markings being nearly absent, and the fibers studded with minute molecules. But in the portions infiltrated with blood the degeneration was more advanced than elsewhere, the striæ were wholly deficient, the fibers crowded, and in some cases opaque, with beads of oil, many of which were of considerable size. The white pus-like spots in the right ventricle consisted simply of muscular fibers extremely degenerated.

"'The kidneys, though looking healthy to the naked eye, were really much diseased. The Malphigian bodies were generally healthy, but a few presented accumulations of oily granules between the capsule, and contained tufts of vessels. The epithelium of the tubes was generally opaque and granular. In many instances the peripheral surface of the cylinder of cells presented numerous oily globules; and not infrequently the tubes appeared filled with separated and irregularly clustered epithelial cells, loaded with oil so as to be almost opaque. In a few cases, tubes were filled with recently extravasated blood; and occasionally transparent casts were seen floating about the field of the microscope. The contents of the medullary tubules were more generally unhealthy even than those of the cortical ones. Many contained transparent fibrinous casts, and the majority presented oily, breaking down, epithelial contents.'

"I am indebted to Dr. Bristowe for the following report of a case, which recently proved fatal in St. Thomas's Hospital. I had not the opportunity of seeing the patient during life, but carefully examined the affected organs after death.

"E. T., a girl, aged eleven years, suffering from club-foot, had been in St. Thomas's Hospital, under Mr. Solly's treatment, since May 22, 1860. On the evening of June 23d, she first complained of sore throat. This increased in severity during the next few days; pain and difficulty of swallowing came on, and on the afternoon of the 27th she was placed under the care of Dr. Bristowe. There had been no marked febrile symptoms, no shivering, headache, or pains in the limbs. Neither in the ward nor among the child's friends had there been any cases of scarlet fever or diptheria; but a little girl in an adjoining bed had been attacked, much about the same time, with a sore throat, which had disappeared in a day or two, and presented no unusual character.

"'*June* 27.—Is perfectly sensible and composed, having by no means the aspect of a person seriously ill. Has no headache, or pains about the limbs; complains of a little thirst and loss of appetite, but no sickness, cough, or difficulty of breathing. Pulse 124. The pupils are natural. The skin is warm, but not dry, and without trace of rash. The external fauces on the right side are much swollen, very tense and tender, but not discolored. On looking into the throat, the right tonsil is seen to be so much enlarged as to appear almost to close the passage, and is covered in nearly its whole extent by a thick, grayish, false membrane. The uvula is pushed over to the left side, and almost concealed; is somewhat thickened, and a little false membrane adheres to it. The left tonsil is hidden, and apparently not enlarged. The tongue is covered with a whity-brown fur, and its papillæ are not prominent.

Hirudines ij. faucibus externis. Catapl. lini postea.
℞ Chlorat. potass.   gr. iv.
  Acid. hydrochl.   ℳj.
  Aquæ dist.   ℥ ss.
      4tis horis.

"'Milk diet. Strong beef tea. Two eggs. Wine, three glasses.

"'28th.—Passed a comfortable night, and has taken all her wine and nourishment. The leeches have given her great relief. There is little appreciable change in either her general health or in the condition of the throat, except that the right side is less tense and tender than it was. The bowels are confined.

Wine, 4 glasses.
Pulv. rhæi c. hydrarg. Ɔj. statim.

"'29th.—Was very restless during the night. The bowels have been relieved, and she has been very sick. The skin is hot, and rather dry. No rash. Pulse 128. No pains anywhere excepting in the throat; no cough or difficulty of breathing. Great pain and difficulty of swallowing. There is copious discharge from the nostrils. Tongue clean. The right side of the throat is in the same condition as yesterday; but the left side also is now distinctly swelled and painful. The right tonsil is about as large as it was; but the membrane, which is thick and tough, is detached and curled up at the margins. The left tonsil is somewhat increased in size, and also presents a distinct false membrane. The uvula is seen with difficulty, but has a few patches on its surface. The lungs are resonant in front; but the respiratory sounds are masked by the noise produced in the throat. Urine albuminous. Sp. gr. 1,015. Wine, twelve glasses.

"'Toward the evening she grew considerably worse, and became very restless. The pulse rose to 152; a troublesome cough, at times a little croupy in character, came on; the breathing became rapid (40 in the minute), and more noisy than it had been. She continued perfectly sensible.

"'30th, nine A.M.—Has been very restless all night, and has taken very little wine and nourishment in consequence of inability and disinclination to swallow. Is now manifestly sinking; is scarcely sensible, but can be roused; breathing rapid, accompanied by loud rattle and frequent moans; pulse imperceptible; lips dry. Died at ten A.M.

"'*Autopsy.*—The body was in a fair condition. There were no traces of eruption or of desquamation. The right submaxillary region was much swelled and indurated; the left also, and the intervening parts, were swelled, though in a less degree.

"'*Chest.*—Pericardium healthy. Heart of natural size, and for the most part healthy. Its external surface presented numerous petechial spots, and its cavities contained partly decolorized coagula. The pleuræ were free from adhesions, and the upper lobe of the left lung was covered by a very thin film of recent granular lymph. The lungs were rather large, heavier than natural, and presented, when handled, the irregularly solidified character distinctive of lobular pneumonia. On section, the upper lobes of both lungs were found to furnish well-marked specimens of the condition just named. They were studded thickly with smallish solid masses, running to some extent into one another, and separated by an imperfect net-work of still crepitant, though congested, lung tissue. The solid masses varied in character; in some instances were distinctly apoplectic, in others had the appearance of being due to simple carnification, and in others presented various degrees of the brick-red tint and granular condition belonging to red hepatization. The lower lobes were, in many respects, in the same condition as the upper; but they presented a greater degree of simple collapse, and, consequently, a less amount of crepitant tissue; the hepatized and apoplectic patches, too, were larger, and presented less of the lobular arrangement. The bronchial tubes were congested, and contained much frothy mucus.

"'The larynx, trachea, and adjacent parts were now removed and examined. The right tonsil was found to be very large, though scarcely so large as during life; the left also was enlarged, but in a less degree than its fellow; and the uvula and soft palate were somewhat thickened. The tonsils, soft palate, uvula, base of tongue, and posterior and lateral part of pharynx were covered, more or less completely, with tough, somewhat elastic, whitish false membrane. On the base of the tongue and uvula it formed merely thin, scattered patches. But over the tonsils, pillars of the fauces, and rest of the pharynx, it formed layers of considerable extent, and often more than half a line thick. The membrane had become generally more or less detached at the edges; and that portion connected with the right tonsil had separated in nearly its whole extent, and hung as a loose, discolored mass, backward into the pharynx. On peeling the membrane off, it was found pretty firmly attached, and accurately molded to the inequalities of the subjacent mucous surface, which was congested, but not ulcerated. On section, the tonsils were seen to be deeply congested throughout, somewhat softened, and studded thickly with small patches of yellowish (but not distinctly purulent) inflammatory deposit. The tissue of the soft palate and uvula was a little brawny.

"'The mucous membrane of the upper part of the larynx was congested and somewhat thickened; and a thin false membrane covered the epiglottis, extended into the aryteno-epiglottidean folds, and down to the superior vocal cords. False membrane also extended into the *sacculi laryngis*, and was scattered in small patches over the mucous membrane for about an inch below. The greater part of the trachea was healthy.

"'*Abdoinen.*—Peritoneum healthy. Liver healthy, but studded with a few pallid patches. Spleen, pancreas, and super-renal capsules healthy. The mucous membrane of the stomach presented numerous petechial spots; and Peyer's patches in the lower three feet of the ileum were remarkably distinct and prominent; in other respects the alimentary canal displayed nothing unusual. The kidneys did not look unhealthy; but exhibited, in their cortical substance, alternate pallid and congested vertical streaks. Uterus and ovaries healthy. Larger blood-vessels natural.

"'*Microscopic Examination.*—The false membrane was identical in its intimate structure with those which I had formerly examined and described. The only unnatural character exhibited by the kidneys was, general great granularity of the epithelium, and consequent opacity of the undenuded tubules. It seemed, too, as though the individual cells were abnormally large. There was no trace of effused blood, and no casts. The Malpighian bodies were normal.'"

DRUG TREATMENT OF DIPTHERIA.

With all the data before us which careful observation, extensive experience, keen analysis, history, mortuary statistics, and morbid anatomy can furnish, we now approach the really important and responsible part of our subject—the treatment of diptheria. All persons who will carefully read the history of all the wide-spread epidemics which have prevailed in the world—the plague, the sweating sickness, the influenza, the scarlatina, the cholera, and the diptheria—can not fail to notice the wonderful harmony of medical authors in description and diagnosis, and the strange discordance of medical practitioners in their manner of treatment. Physicians who agree precisely as to the seat, character, and causes of the disease, will recommend exactly opposite methods of treatment; while others who disagree as to the type and diathesis of the malady, will agree in their plan of medication. This is not only true of diptheria and other pestilences, but of all diseases. And the explanation is, that the medical profession has a false theory of all diseases—of the nature of disease itself.

The majority of physicians recognize the diptheritic exudation to be an inflammatory process. But what is inflammation? Here all is discord and confusion again. "It is *increased action*, and must be reduced," says one; and in goes the lancet, or on go the leeches, or down go the emetics, the purgatives, the antiphlogistics, etc., and down and off goes the patient.

"Inflammation is *decreased* action," says another, "and the patient must be sustained through it;" and in and down go brandy and quinine, wine and cordials, beef-tea and egg-toddy.

"The inflammation is *specific*," says a third, "and must be specifically counteracted;" and the mucous membrane is seared with lunar caustic, scorched with cayenne pepper, burned with alcohol, excoriated with chlorate of potassa, denuded with sulphate of zinc, corroded with hydrochloric acid, and constringed with preparations of iron.

"The local inflammation is *active*," is the doctrine of another; and nitrate of potassa and antimonial wine are the remedies.

"The local inflammation is *passive*," exclaims another; and aqua ammonia, and mustard poultices, and turpentine liniments, and alcoholic gargles are in requisition.

"The exudation is a *parasitic fungus*," replies another; and death to the animalcules is dealt out in the shape of calomel, nitrate of silver, anguintum, sulphur, arsenic, iodine, iron, salt, alum, etc.

"But the chief difficulty lies further back; it is a *blood disease*," says another; and so he attacks the virus by sending an antidote, a counter-poison, a drug, a medicine into the blood, in the vain expectation that, in some mysterious manner the poison he sends into the system will neutralize or destroy a worse poison. Says Prof. Jos. M. Smith, M.D., of the New York College of Physicians and Surgeons: "All medicines which enter the circulation *poison* the *blood*, in the same manner as do the poisons that produce disease."

"The disease is essentially a *fever*, and requires the alterative and evacuant plan, emetics, cathartics, diaphoretics, etc.," is the teaching of another.

"The fever is *sthenic*," exclaims another; and digitalis, antimony, niter, and acetate of ammonia are prescribed.

"The fever is typhoid," replies another; and the patient is stimulated through the whole course of the disease, and perhaps for months, if not years, after.

All the books which have thus far been written on diptheria have recommended some form or modification of drug-medication; and as I am writing one against drug-medication, and in favor of hygienic treatment, and as I wish to turn the public judgment as much as possible against drug treatment of every kind, I know not how I can better accomplish this object than by showing precisely what drug treatment is, according to the most approved authorities, and what the testimony of the different practitioners is, respecting the effects of the treatment as recommended by their professional brethren. If the reader does not see, in this exposé, ample reason for discarding all drug-medication, and relying on hygienic agencies alone, "neither could he be convinced though one should rise from the dead." It will at least prepare him the better to appreciate the rationale of hygienic medication; and I think this exposition will enable the candid mind to understand the why and wherefore of much of the mortality of diptheria, and of many of its complications and sequelæ.

Dr. Slade, who regards diptheria as a *specific* disease, propagated by infection and contagion, and belonging to the category of *blood diseases*, remarks, in relation to the methods of treatment formerly in vogue:

"Like all diseases which have prevailed epidemically, and which have appalled by their severity and fatality, or perplexed by their novelty, diptheria has been subjected to a great variety of treatment. It is only within the last four years that anything like a unanimity has existed in the profession in regard to this important point. Not to go farther back than the period of Bretonneau's memoir on this subject, we shall find that an activity of treatment prevailed which would scarcely coincide with the ideas of the present day."

What is "activity of treatment?" If this phrase has any meaning at all, it means *killing*. And we shall all be glad to know that such treatment does not coincide with the ideas of the present day. But I fear it is the result of the practice of the present day, and even of that practice which Dr. Slade recommends. Dr. Slade continues:

"Bleeding, both local and general, blisters, certain local applications to the pharynx, rapid mercurialization, formed the treatment in all cases. Mercury, in fact, was considered as the sheet-anchor by a great majority of medical men. To quote the words of Dr. Samuel Bard: 'But, although I consider mercury as the basis of the cure, especially in the beginning of the disease, I do not by any means intend to condemn or omit the use of proper alexipharmics and antiseptics.' Although a few practitioners may still make use of this therapeutic agent, it is now generally agreed that such is the asthenic nature of the disease at the present day, that depletion is not borne well in any form, neither is the action of mercury defensible either in theory or practice."

If depletion can not be borne in diptheria, it is because the patient sinks under it; and if the use of mercury is not defensible, it is because it damages or kills the patient. That such are the results of these "therapeutic" agents, Dr. Slade testifies, though in a very gingerly manner; and it is a sad pity that the profession can not see that this truth applies to all other febrile and inflammatory diseases—and even to venereal diseases, for which it is claimed to be *the* "specific remedy"—as well as to diptheria. Says Dr. Slade again:

"As we are not yet acquainted with any specific capable of arresting the course of diptheria, our treatment must be directed simply to the conducting our patient in his progress through the disease."

Think, reader, seriously, for one moment, of the idea

of a patient being *conducted through a disease!* More frequently still we hear of a disease "running its course" through the patient. Is it not about time that the profession settled the question, whether the patient passes through disease or disease passes through the patient? But such expressions, it will be claimed, are not literal but figurative. They are literal nonsense and figurative foolishness. They indicate, as well as language can, the "incoherent expressions of incoherent ideas," which constitute the chief burden of medical literature, so far as the nature of disease and the action of remedies are concerned. When medical writers learn the simple truth, that disease is not a thing which runs through living organisms, nor which can be run through by a person, they will cease to employ such senseless and unmeaning language. And when they understand that disease is vital action in relation to things abnormal, they will see a better way to treat it than by the administration of drug-medicines.

Dr. Slade objects to blisters, because their irritation aggravates the engorgement and cellular infiltration, and also because the blistered surface is liable to put on a diptheritic or sloughy appearance; and he condemns bleeding, "except, perhaps, in very rare exceptional cases." Emetics, he thinks, may be advisable "under certain circumstances," and when there is a tendency to croupal symptoms; and then he recommends full doses of ipecac. He condemns "anything like purging," but approves simple enemas and mild laxatives.

"There are occasional cases of diptheria so mild in character that local applications to the fauces may be sufficient; but as a general rule it may be conceded that the disease requires a tonic and sustaining treatment; particularly is this often the case at a late period of the disease."

Can it be possible that the *disease* requires a tonic and sustaining treatment? So says Daniel Dennison Slade, M.D., of Boston, Mass.; and Dr. Slade received a premium of one hundred dollars awarded by the Trustees of the Fiske Fund, at their annual meeting held at Newport, R. I., July 11, 1860, said Trustees consisting of James H. Eldridge, M.D., of East Greenwich, Charles W. Parsons, M.D., of Providence, and Henry D. Turner, M.D., of Newport—all of which facts are attested by S. Aug. Arnold, M.D., of Providence, the Secretary of the Fiske Fund—for the Essay which contains this somewhat startling announcement. And it may be pertinent also to remark, in this place, that this Prize Essay which contains this remarkable statement, was published in the *American Journal of the Medical Sciences* for January, 1861; from which it has been reprinted in book form for consultation and reference.

Were it not that we are dealing with a Prize Essay, indorsed by the Rhode Island Medical Society, we might be disposed to criticise the idea and dispute the propriety of *sustaining the diptheria* with tonic treatment. If anything requires tonic and sustaining treatment, it seems to me it is the *patient*, and not the disease. But as Dr. Slade says it is the disease which requires, and as the "authorities" are all on his side, I suppose we shall have to submit, which I do under the protest that I can not comprehend the matter at all. And the author has still further complicated the matter, in representing that the *disease* requires the tonic and sustaining treatment particularly at a late period of the *disorder;* in other words, diptheria should be sustained by tonics in a late period of diptheria! Why not let the disease run down and die if it will? I can imagine no method for sustaining the disease except to add to its causes, and that would be *feeding the diptheria* sure enough! But I am of opinion that we had better feed the patient—add to his causes—and let the diptheria go.

But, after all, Dr. Slade has employed none but recognized medical parlance. There is not an author of a text-book on The Practice of Medicine who does not frequently use language in the same sense, or the same nonsense, and who does not habitually confound causes of diseases, consequences of diseases, the actions of disease, the disease itself, and the patient. Nor will they, nor can they ever avoid this confusion worse confounded until they get a new and a true theory of the nature of disease, and of its relation to the vital organism. Says Dr. Slade:

"Stimulants and nourishment should be commenced with early, and persisted in systematically. The amount, of course, must depend upon circumstances; but in order to insure efficiency, they should be varied, should be given in small doses at regular and frequent intervals, and if rejected by the stomach should be given in the form of enemata. So also with respect to children, when they are frightened and disturbed by painful attempts at swallowing, and absolutely refuse everything, we have the same resource: Injections of beef-tea, with brandy and quinine, may be employed, and thus life may be not unfrequently sustained, when otherwise it would inevitably have been extinguished."

As it is "life" now, and not disease that is to be sustained, I go for the principle, but do not like Dr. Slade's manner of applying it. I must infer from his medico-alimentary medley that he is still a little muddled as to what he ought to prescribe for. Beef-tea is poor nourishment for diptheria; and brandy and quinine are wretched food for the patient. Beef, being food, may contribute to the life of the patient; while brandy and quinine, being poisons, must inevitably add to the causes of disease. In almost every instance in which we have known a practitioner to get inextricably befogged between conflicting theories, or to become perplexingly embarrassed with "indications and contra-indications," or to be grievously harassed with doubt whether he ought to give one set of remedies or just the opposite, he has solved the difficulty by a compromise, adopting a little of each of the theories, and mixing up some of both kinds of remedies. And this seems to have been the case with Dr. Slade.

"With regard to the particular form of tonics," says Dr. Slade, "there is a variety of opinion. There are some which, perhaps, promise a greater chance of success than others, among which we may mention quinine, tincture of chloride of iron, and chlorate of potash. But as each of these has powerful advocates in its favor, we imagine that, provided the strength of the patient be sustained, it is of little importance by which of these tonics it is accomplished."

"*Provided?*" But there's the rub. *If* one poison *will* sustain the strength of the patient—and we ought to be thankful to know distinctly that it is the patient and not the disease, the "*strength* of the patient" that should be sustained—it is not, of course, of so very much importance what other poisons are administered or withheld, whether their advocates be powerful or weak.

After indicating his preference for the tincture of

sesqui-chloride of iron—as the best of the many internal remedies which have been advised—with chlorate of potash, chloric ether, and hydrochloric acid in the form of mixture, sweetened with syrup, and given in full and frequent doses, Dr. Slade quotes approvingly from the *Lancet* the following remarkable passage :

" A free use should be made of generous wine, beef-tea, coffee, eggs, in combination with brandy and wine, milk, and whatever other form of nutriment the ingenuity of the surgeon or the fancy of the patient can suggest."

A more horrid jumble of dietetic druggery, or medicated food, can scarcely be imagined; but if surgical ingenuity, constructive or destructive, or invalid fancy, normal or morbid, can suggest anything else, by all means let the patient have it! Is this the medical science of the nineteenth century? Is this the healing art of A.D. 1862? And is a prescription of diet and drugs, separately or in combination, a *surgical process*, to be devised by the " ingenuity of the surgeon?" It is not even a chemical combination, but a mechanical admixture of pathological and alimentary ingredients, anti-pharmacologically compounded, and most unphysiologically confounded. Were not our subject a grave one, we should be disposed to wield no weapon but that of ridicule against such superlative nonsense.

But, seriously, we protest against stuffing the stomach with anything, much less with these incongruous abominations. The patient can not digest food of any kind during the acute stage of the local inflammation, nor until the violence of the fever has subsided; and to burden the system with anything which it can not use, under the circumstances, is merely to nourish and sustain the *disease* by adding to its causes. When the vital powers are wholly occupied in a life-and-death struggle, as it were, to expel impurities from the machinery of life, or to deterge a virus from the blood, they can digest nothing; and to gorge the digestive apparatus with a promiscuous medley of slops and stimulants, is to sustain the disease and exhaust the vitality.

I dwell on this point with some emphasis; for there is no greater delusion in the world than that which mistakes *stimulation* for *nutrition*. The ideas are exactly antagonistical; and yet the whole medical profession has for ages, with less than one exception in a thousand, prescribed stimulants to support the vital powers, when the digestive function was feeble or suspended, as though stimulation was the equivalent of or a substitute for nutrition. Instead of supporting vitality, stimulants, of all kinds, exhaust it; they occasion its preternatural expenditure, and all such *use* of vitality is *abuse;* it is waste, and nothing else; as is all abnormal action under all circumstances. Need any one wonder at the grave complications, the numerous sequelæ, and the many and serious cases of paralyzed muscles and prolonged convalescence, in view of such methods of treating diptheria, or drugging the patient?

We have now done with the general and leading remedies which Dr. Slade recommends to be administered to patients suffering of diptheria, and we come next to the local and auxiliary measures; and as the Prize Essay of Dr. Slade is confessedly " a full and accurate resumé of what is known concerning a disease which is now attracting universal attention," and is a fair compendium of the views and practices of the American medical profession, I propose to examine it somewhat critically to the end.

" We come now to speak of the auxiliary measures

to be adopted in the treatment of this disease, and first, of the local applications to the fauces. The propriety of these has been called in question by some writers, on the ground that the disease is a constitutional one, and, therefore, that they can be of no service. But we must answer to this, that there can be no more reason why the local remedies are not as applicable to this affection as in other constitutional diseases, for example, as in syphilis, scrofula, carbuncle, etc."

Dr. Slade next indorses as "excellent" the following reasons given by Dr. Bristowe for discarding heroic applications to the fauces:

"1. That the throat affection is merely a local evidence of a constitutional disease, which is unlikely to be arrested in its progress by any treatment directed to the secondary manifestations only. 2. That the throat affection rarely kills, except by involving organs, such as the trachea and deeper tissues of the neck, which are beyond the reach of the possible influence of such agents. 3. That if the theoretical correctness even of such treatment be admitted, the application of remedies to the surface of a thick false membrane, with the hope that they may affect the subjacent mucous tissue, is not only clumsy, but, as regards the object intended, practically useless; and that the prior forcible removal of the membrane from the entire surface, in order to their efficient employment, is unjustifiable in the early stage, even if possible, and is likely only to be followed by increased inflammation, and production of false membrane."

Nevertheless Dr. Slade is for a compromise. He says: "While we concur in the remarks of Dr. Bristowe so far as regards the forcible removal of the membrane, particularly in the early stages, the experience of almost all medical men of the present day bears witness to the efficacy of the application of caustics or escharotics to the throat."

We think very little of the experience of medical men, who, in adopting a false theory of the nature of disease, must necessarily interpret the effects of remedies by erroneous standards. Experience informs us what medical men *have* done, not what they *should do*. And we shall see, presently, that some practitioners of great experience declare that caustics only aggravate the disease and extend the local inflammation; they give a reason, too, why escharotics should not be employed in any stage of diptheria; and I can not help having more respect for one sound reason, one true principle, one demonstrated theory, than for all the experience of all the medical men of all the world in all the ages, so far as that experience is judged by the false standard of the prevalent medical doctrines. Dr. Slade continues:

"On the other hand, some writers maintain that the disease at the outset is a local one, which rapidly brings on a general *intoxication*. This would be a still stronger argument—if we granted this to be true —for these very local remedies, if applied in season, might prevent a further extension of the disease."

Surely the profession is in a most unfortunate predicament—unfortunate at least for the patient—with regard to the rules of practice by which they should be governed in the treatment of diptheria. In the first place, the authors do not agree whether the disease is general or local; nor, if constitutional, whether general or local remedies are to be put most prominently forward in its treatment. But the reasoning of Dr. Bristowe, that the trachea and deeper tissues of the neck "are beyond the region of the possible influence of such agents," I hold to be entirely and mischievously fallacious. Any poisonous agent in contact with any part of the living system, influences, to some extent, every organ and structure. Its presence invariably occasions some disturbance in the part or organ to which it is applied, and a less degree of disturbance in organs and structures more remote; just as the presence of a thief in the family circle, or of a serpent in a promiscuous crowd, would occasion a general commotion among all the persons present, and a greater consternation or resistance among those in contact with or nearest to the offending thing. It is true, the effect or influence of a poison on a part distant from the point of contact, is not always appreciable, nor is its local influence always apparent; yet, if we understand the law of constitution and relation between dead and living matter, we know that, whether cognizable to our senses or not, some effect must result, just as we know that when we add a drop of water to the Croton Reservoir, the bulk of the whole mass of fluid is increased, although our eyes can see no difference. The constant dripping of the soft water will in time wear away the solid rock; yet our eyes could recognize no change from day to day; and so the constant use of stimulants, irritants, nervines, narcotics, and, indeed, of any other drug or poison, gradually and imperceptibly exhausts the life-power, until the accumulated debility brings us to the recognition of the law of vitality, and the consequences of abnormal vital expenditure.

It is quite common for medical men to say, when their remedies have not benefited the patient, that they have had no effect whatever. This is impossible.

They do and must occasion vital action. Nothing can be inert or neutral in relation to the living organism. It is either useful or injurious; and its administration as a medicine, as well as its accidental presence, must and does always, and under any circumstance, exercise, so to speak, an influence. Of the individual agents employed as local applications, Dr. Slade testifies very dubiously and rather ambiguously:

"There are a multitude of substances which have been employed as local applications to the fauces, all of which have their special advocates. During the last four years, the nitrate of silver, either solid or in solution, has been perhaps more extensively used than any other substance. This, when used early in the disease, seems in many cases to check the progress of the exudation; yet it does not answer the purpose altogether, and further experience has somewhat diminished confidence in it. Indeed, in some instances it is a question whether the free application of this caustic does not rather add to the evil."

Nitrate of silver, we are told, has been more extensively employed than any other caustic, and experience diminishes confidence in it. What are we to do? Dr. Slade has told us, a little way back, that "the experience of almost all medical men of the present day bears witness to the efficacy of the application of caustics or escharotics to the throat." The testimony is in favor of *some* cauterizing agent, and against nitrate of silver. What, then, is the proper article? Dr. Slade solves the difficulty in his usual half-and-half compromising manner:

"Still, if carefully and properly used, nitrate of silver in many cases is undoubtedly of benefit. If in solution, it is to be applied by means of a probang or brush, swabbing over the diseased surface quickly, at the same time thoroughly. The strength of the solution should be from 30 to 60 grains, and perhaps higher, to the ounce of water, according to circumstances. For children, a full-sized camel's hair brush is best. The child should be placed on the lap of an attendant, and the head firmly fixed. If he will not open the mouth, the nostrils should be closed for a few moments, and as he opens the mouth for breath, the jaw should be at once depressed, and then, the tongue being kept down by the finger, the fauces are brought well into view, and the solution thus thoroughly applied. The utmost gentleness and patience should be exercised; at the same time, firmness, for upon the effectual accomplishment of this proceeding the success of the treatment will greatly depend. This should be repeated every three or four hours, so long as it is necessary."

"According to circumstances," "and so long as it is necessary," are rather loose and indefinite rules for the application of so potent an agent, especially when we are given to understand that the profession is divided on the question whether it tends to cure the disease or to kill the patient; and more especially when our author is himself partly on both sides of the question, and partly between the rival opinions. Nevertheless, Dr. Slade, with commendable candor, quotes a brief chapter of the evils of cauterization, from the pen of F. A. Bulley, F. R. C. S., and the *Medical Times and Gazette* for April, 1859:

"I have mentioned that I thought that the indiscriminate mopping of the fauces, as it is called, with solutions of nitrate of silver, was frequently attended with injurious results in this disease, principally, I believe. for this reason, that, owing to the struggles of the little patient, it is impossible to apply the caustic solution with that precision which the case absolutely requires. Thus, it is applied to parts which are entirely free from disease. I have been told of cases where the inside of the cheeks have been covered with it; in coughing, a portion of it has been expelled upward through the nose, corroding the susceptible surface of its mucous membrane; and again, other portions of it have seemed to pass downward into the pharynx and esophagus; and I am not sure that, during the convulsive struggling of the patient in resistance, some of it may not also enter the larynx, where it may possibly initiate those inflammatory changes in the mucous membrane of the air-passages, which are too frequently the harbinger of death in this disease."

To the adverse testimony of Dr. Bulley, Dr. Slade adds the following *pro* and *con:*

"The nitrate of silver may also be employed in the solid form, but this we should not advise, particularly in the case of children. During the struggles of the little patient the crayon might become broken, an accident which has happened, and fragments fall into the esophagus or larynx, giving rise to serious lesions. Moreover, the nitrate of silver in this form has the disadvantage of creating a more decided eschar than does the solution, simulating the diptheritic exudation, and thus hindering the perception of the progress of the disease."

Dr. Slade regards the tincture of chloride of iron as an excellent substitute for the nitrate of silver, and commends hydrochloric acid "in some cases," but forgets to tell us to what cases it is adapted. In the case of children, the addition of honey to the acid is recommended. Hydrochloric acid was a favorite topical application with M. Bretonneau, who preferred to employ the agent in its full strength, at long intervals, than to return to less energetic applications more frequently.

Dr. Slade also mentions commendatorially as local applications, a solution of the chloride of soda, chlorate of potash, and the combination of chlorate of potash and hydrochloric acid with the tincture of the sesqui-chloride of iron, this combination being especially adapted to croupal cases; and the chlorate of potash, we are informed, has "an undoubtedly anti-diptheritic influence, where time exists to bring it into play."

This *anti-diseaseical* influence of a remedy reminds me of some of the celebrated preventive horse-medicines of Dr. Dadd, the veterinary surgeon, which, to borrow his beautifully philosophical expression, operarate *anti-pathologically.*

Among the numerous other applications to the fauces which are employed and recommended by practitioners, Dr. Slade mentions strong solutions of sulphate of copper, chloride of sodium, tannin, capsicum, and Monsell's salt. With regard to the virtues of Monsell's salt Dr. Slade quotes the following testimony of Dr. Beardsley, of Milford, Conn.:

"Monsell's salt was found to be the most efficacious and valuable of all topical remedies, affording in some instances decided relief. Its active astringent property rendered it peculiarly appropriate, and well adapted to obviate that relaxed and enfeebled condition of the throat which attends the advanced stage of the disease."

We have already seen that fourteen out of Dr.

Beardsley's fifteen cases terminated fatally, and that the fifteenth case was probably saved by running away from the doctor; and in view of these facts, and of the statement of Dr. Slade in relation to these cases, viz., "there was nothing peculiar in the treatment," the opinion of Dr. Beardsley that Monsell's salt "was found to be the *most efficacious* and valuable of all topical remedies," affording, *in some instances*, "decided relief," its astringent properties being "peculiarly appropriate," etc., must be taken for what they are worth. The efficacy, so far as results were concerned, seems to have been in the wrong direction.

In cases where there is much *tonsillitis*, the inhalation of steam, mucilaginous gargles, and warm fomentations are recommended; and M. Bouchat has advised the *removal of the tonsils* in the early stage of the disease. To this tonsillitic ablation Dr. Slade raises the following objections: "In the first place, the exudation is almost sure to re-form upon the cut surface; next, there is a great risk of severe hemorrhage; and finally, any cutting operation, however simple, had better be avoided, if possible, especially upon young children, and in a disease so asthenic in its character."

For the purpose of facilitating respiration in an adult, in cases of great tumefaction, the removal of the tonsils, Dr. Slade says, "might possibly be practiced," by which expression I presume he means, *might possibly be justifiable.*

When the nasal fossæ have become implicated, various solutions and powders have been recommended to be employed by injection and by insufflation. MM. Bretonneau and Trousseau preferred alum. Dr. Slade advises chloride of soda and glycerine; also frequent injections of warm water and soap as a cleansing process. Dr. Slade adds: " Injections of nitrate of silver, sulphate of zinc, and, in fact, any solution which is applicable for the fauces, will answer a good purpose for injecting the nasal cavities."

All very easy to write. But the practical difficulty is to find whether these things are applicable or not to the fauces; and the testimony of the authors we have thus far quoted, leaves this matter decided *both ways*, and this is what we term proving *too much*, and thereby invalidating the evidence.

Dr. Slade, in conclusion, gives us a summary of the arguments and authorities for and against the operations of *tracheotomy* and *tubing the larynx*, which subjects I shall refer to again.

Having thus reviewed the whole theory and practice of Dr. Slade—whose Prize Essay gives us the substance of the doctrines and prescriptions of the medical profession in relation to diptheria—let us briefly glance at the teachings of other authors and practitioners. And first we turn to the latest author of a standard work on theory and practice (" Wood's Practice of Medicine"), which work is a text-book in our medical schools. Dr. Wood does not agree with Dr. Slade, that the disease is always asthenic, requiring the stimulant and tonic treatment from the first. On the contrary, Dr. Wood regards it as sometimes of the opposite diathesis, and, accordingly, he recommends the very opposite treatment—bleeding, salts, etc. Indeed, the general plan of treatment recommended by Dr. Wood in his standard work, is the very treatment which is condemned by Dr. Slade in his Prize Essay.

As an illustration of the "unanimity" which does *not* prevail in the medical profession respecting the nature and treatment of diptheria, let us place the principles of medication advocated by these distinguished authors in contrast.

Says Dr. Wood (vol. i., p. 553): "In the mildest cases little general treatment is required. The patient may take a dose of sulphate of magnesia, or some other saline cathartic, and should avoid animal food. In somewhat severer cases, with moderate fever, the cathartic may be repeated, and antimonials and the neutral mixture administered at short intervals. When the pulse is full and strong, blood should be taken freely from the arm, especially in adults; but venesection does not exercise the same controlling influence over this as over the common inflammation; at least, it does not obviate the tendency to the plastic affusion; and, in some instances, in consequence of the feebleness of the system, is not well borne. It is generally quite inapplicable to those cases which occur epidemically, or in which a dark hue or fetid odor of the exudation indicates a depraved state of the blood. When the symptoms are threatening, either from the general condition of the system or the disposition in the local disease to enter the respiratory passages, calomel should be resorted to. Under these circumstances, no general means of cure is so effectual as the establishment of the mercurial influence. If the patches should have reached the glottis, or be extended toward it, a full purgative dose of calomel should be given, and the medicine afterward continued in doses of from half a grain to two grains, every hour or two, until the mouth is affected or the disease relieved. Even young children, under these circumstances, bear calomel well in the quantity mentioned. Should it irritate the stomach and bowels very much, the dose may be diminished, or the mercurial pill, and frictions with mercurial ointment, substituted."

If there is anything loose, slip-shod, vague, indefinite, or ambiguous in the Prize Essay of Dr. Slade, it is equaled, if not exceeded, in the "Practice of Medicine" of Dr. Wood, while on the main points of treatment, these authors are diametrically opposed to each other.

Dr. Wood not only recommends the depleting plan in many cases, but assures us that it is the very best. He says: "No general means of cure is so effectual as the establishment of the mercurial influence."

Dr. Slade says: "The action of mercury is defensible neither in theory nor practice."

Dr. Wood says: "When the pulse is full and strong, blood should be taken freely from the arm."

Dr. Slade says: "Depletion is not borne well in any form."

Dr. Wood prescribes "full purgative doses of calomel."

Dr. Slade replies, "Anything like purgatives should be sedulously avoided."

Dr. Wood recommends the most depleting and debilitating drugs of the materia medica—neutral salts and antimony.

Dr. Slade insists that such practice is always injurious.

Well, what is the young practitioner, or the old one, to do, when he goes forth to combat diptheria, with these high authorities in his hands? Probably he, too, will compromise, and adopt partly the practice of each, and so *do* with one hand and *undo* with the other.

Dr. Wood recommends, as external applications to the throat, leeches, rubefacients, and blisters; and, in relation to the internal local applications, he advises: "By far the most important remedies are those addressed immediately to the part affected. By these the peculiar character of the inflammation, upon which its danger chiefly depends, may be changed; and if the disease has not already reached the larynx, its progress may be arrested. In the slighter forms, a solution of sulphate of zinc, containing fifteen or twenty grains of the salt in a fluid ounce, applied daily or twice a day to the pseudo-membranous patches, will be found sufficient. When a stronger impression is required, caustic substances must be employed. Of these the best is nitrate of silver, which may be applied either in the solid state, or dissolved in six or eight parts of water. Muriatic acid is highly recommended by some writers, and in the worst cases is used undiluted. In those of slower progress, it may be diluted more or less according to the impression desired. Alum is another very efficient application. It is used in saturated solution, or in the form of a very fine powder, which is applied directly to the part by blowing it through a tube adapted to the purpose. These substances should be allowed to come in contact as little as possible with any other part of the surface than those covered with the exudation. The liquids may be applied by means of a large camel's hair pencil, or of a piece of sponge or soft linen attached to the end of a stick. In the intervals between the caustic applications, mucilaginous gargles, sweetened or not with honey of roses, may be beneficially used. A gargle made of a fluid dram of chlorinated soda and four fluid ounces of water, is recommended in cases attended with fetid discharge.

"Howard's calomel, applied to the diseased surface by means of a tube, was advised by Bretonneau; but its chief advantages are probably derived from the portion of it which may be swallowed. When the disease enters the nasal passages, the solution of nitrate of silver may be injected up the nostrils."

In an article published in the *Medical Times and Gazette*, of Sept. 3, 1859, by Dr. J. S. Bristowe, of Southwark, the author is entirely opposed to the practice so confidently advised by Dr. Wood. He says:

"An important question is that having reference to the mode of treatment of the affection of the throat; and I may here state, as may have been inferred from the perusal of my cases, that I, for one, disapprove of the application to the diseased surface of strong caustics and escharotics, and should prefer the employment in all cases of mild detergent gargles, or of warm milk, and such like bland and soothing fluids."

We have already quoted the reasons which have led Dr. Bristowe to discard heroic applications.

Dr. C. Swaby Smith, of Barbage, Wiltshire, gives his experience in the London *Lancet* of Sept. 10, 1859:
"I have tried many modes of treatment, and so far with very good results; but the one that I have most faith in is one that I would advise those who have not used it, at any rate just to give it a trial. On first seeing my patient, I apply a strong solution of chlorinated soda to the fauces; and follow up my treatment by ordering a sinapism to the throat; a gargle, composed of solution of chlorinated soda, two ounces; tincture of myrrh, two drams; water, six ounces; to be used every half hour; and in cases where the children are too young to gargle, I order the throat to be frequently washed with the same mixture by means of a piece of sponge. Internally, I give to an adult (of course varying the dose according to my patient's age),

chlorate of potash, two drams; diluted nitric acid, three drams; solution of cinchona (Battley's), one dram; water, to six ounces; the sixth part to be taken every two hours. And in cases where there is much pain in the limbs, I generally add a few minims of tincture of colchicum; which addition has proved decidedly advantageous; the diet to consist of strong beef-tea, port wine, and, in short, all the nourishment the patient can take."

Dr. Smith is, of course, a believer in the absurd "respiratory food" theory of Liebig and others, or he would not conjoin an alcoholic stimulant with a solution of beef, under the head of diet. But I protest against his rule regulating the quantity of diet—"all the patient can take." The patient might be able to take a gallon a day, when he could not digest more than a pint. The rule for the administration of food should be, in all cases of diptheria, whatever the patient can *use*. Food is only beneficial as it is assimilated, not according to the quantity swallowed.

Dr. Smith's concluding remark casts a shade of suspicion over the supposed beneficial effects of his pungent gargles, etc. "Although these means are undoubtedly useful in decided cases of malignant sore throat, they are far too active to be resorted to in simple cases, as they would only tend to aggravate the symptoms."

My own explanation is this: These "active means," or strong applications, which aggravate the symptoms in mild cases, do not benefit the malignant cases; but because of the less degree of vital resistance to the drugs in the malignant cases they *seem* to be well borne, and the practitioner is deluded into the notion that they are useful.

Prof. Clark condemns, in the strongest terms, the depleting and cauterizing treatment so strongly recommended by Prof. Wood and others, and relies almost entirely on the stimulating plan; and his summary of the conflicting opinions of various authors and practitioners, if it does not prove how diptheria ought to be treated, at least shows how little reliance can be placed on medical experience and testimony. I quote entire what he says of the treatment of this disease in concluding his lectures on the subject:

"There is no established treatment for diptheria. In saying this, I speak of the whole treatment, for I think one rule in the management of diptheria is as well established and as generally insisted on as any in medicine—that is, to sustain the patient's strength by food, tonics, and often by stimulants, during the whole course of the disease, and to do this in the face of every difficulty. This rule is not an arbitrary one, but is the result of an extended and almost uniform experience in Europe and in this country. Many physicians, in their early acquaintance with the disease, have adopted the opposite plan, but have found that bleeding and depressing agents generally could not be safely persisted in. Bretonneau, in his second memoir, read before the Academy of Medicine in Paris in 1821, is very explicit on this point. He says (p. 4): 'With regard to *Epidemic Croup*, I am compelled to declare, contrary to the generally received principle, that abstraction of blood has appeared to me hurtful, and to accelerate the propagation of diptheritic inflammation. Emetics and blisters have been used without relief; and I can assert that these means have not been omitted in the greater number of patients who have died.' 'I have not abandoned it [depletion] without hesitation (though it was condemned by the physicians of the seventeenth century); I have been compelled, nevertheless, to yield to evidence, seeing so frequently the opposite of that which I had hoped. I am certain that the symptoms of croup [tracheal diptheria], so far from being retarded, have several times manifested themselves immediately after the application of leeches, applied for the purpose of preventing this fatal disease, the fear of which had been excited by a very slight sore throat. I am now astonished that I did not sooner understand that sinapisms, pediluvia, and irritant injections were measures which were not appropriate to the nature of the disease, and were without proportion to its severity.'

"Dr. Turner, of Petersburgh, Va., has given us his experience with this plan of treatment (*Am. Med. Times*, Dec. 8, 1860): 'Depletants, mercurial alteratives, leeches, blisters, caustics, and common sage gargle, constituted my treatment when I first encountered diptheria.' 'Those patients in whose treatment I employed mercury and local depletants fared the worst.' 'I soon determined that the disease was ultimately *asthenic*, and from this fact I derived the basis of what I consider sound treatment.' In this experience of M. Bretonneau and Dr. Turner you have an account of what has occurred in the observation of many a sound practitioner, and of what will happen to you, I doubt not, unless you begin where such men end, in an entire abstinence from depletory measures, whether general or local. You have but little temptation from the examples of American physicians to abstract blood, because we had been apprised of what our European brethren had learned about it, long before the disease reached us. But knowing what power is ascribed to blood-letting in the management of inflammations, you would be almost forced to a trial of it unless you are informed how worse than useless it has been found by those who have preceded you. We may say, then, that general sanguineous depletion is forbidden in diptheria, and if local bleeding is ever admissible it is only in exceptional cases. I will give you one quotation more in support of this statement from one of the high authorities on this disease. Trosseau (Mems., p. 241) says: 'If diptherite did not differ from simple inflammations in its form, its progress, its dangers, and, in a word, in numerous characters which make it altogether a special disease, it might be supposed that antiphlogistic treatment would be serviceable; but we may conceive, *à priori*, that blood-letting and emollients would have no favorable influence, and experience has confirmed what analogy had led us to surmise. It is in vain to attempt to cure diptherite by means of the antiphlogistic regimen. The inflammatory complications may be subdued, and yet the disease remains without losing any of its malignity.' If we can not inherit the wisdom of those who precede us, we can at least profit by their learning. It is for that reason I have taken time to place this important point fully before you.

"Diptheria is not a 'self-limited disease,' in the sense in which scarlet fever, measles, and small-pox are said to be self-limited, yet it has a duration, varying much in different cases, but which rarely exceeds twenty days for the membranous and most dangerous period. If death does not occur in three, five, seven, ten, or twenty days in the different varieties and forms of the disease, we look for recovery. If we can sustain our patient through these trying periods, however varying, we have done much to insure his recovery. The virulence of the disease has exhausted itself, or at least its power to destroy is greatly diminished. What an eminent medical writer has said of typhus, can, I think, with equal truth be asserted of the constitutional management of diptheria: 'Our treatment can only be of benefit in an indirect manner, that is, in concert with the salutary efforts of the vital powers.' Excluding, for the present, considerations relating to local applications, I may go further and say of this disease what Dr. Stokes says of fever: 'We can not cure fever. No man ever cured fever. It will often cure itself. * * * We prevent dying of exhaustion by food, by the use of stimulants and tonics. * * * We seek to preserve the patient at the least expense to his constitution up to the time when, by natural laws, the disease will spontaneously subside.' Here for 'fever,' read *diptheria;* transpose the words 'stimulants and tonics,' so as to give the higher position to the latter; then, even without reservation in favor of local applications, I believe you have found the great fundamental fact in the treatment of diptheria. I do not wish to say, however, that the rigid enforcement of this important rule for twenty days will always be sufficient, or to say that in every case the dangers are all passed in that time. I have known the death of a child to occur thirty days after the first appearance of the membrane in the throat, and fully three weeks after the exudation in the larynx and trachea had been fully cleared away. Yet this I believe is but one case in a hundred. In general, among the very worst cases, those who have passed the dangers of the first three weeks recover. But the rule of treatment is applicable with modifications to the cachexia which often follows the bad cases, and to the paralytic affections which, though they are but little dangerous to life, are prolonged for weeks and sometimes for months. Having impressed, as I hope, these leading doctrines upon your minds, that blood-letting, both general and local, has been tried in vain; that active cathartics do no good; that emetics are worse than useless, except for a special purpose to be mentioned by-and-by; that revulsives can accomplish nothing advantageous; in a word, that debilitating treatment but plays into the hands of the disease, if I may be pardoned the expression; and that all perturbating general treatment is forbidden; but that food, tonics, sleep, and stimulants (when needed) are the true antagonistics of diptheria—we will now try to appreciate the value of the local treatment, and determine, if we can, whether we possess any agents which have power to prevent or control its justly dreaded local manifestations.

"Here, if I yield to my own convictions, I must say we pass from the certain to the doubtful. It is with reference to these local applications that we are compelled to say that diptheria has no established treatment. If we ask whether nitrate of silver, muriatic acid, or any other caustic can stop the progress of this membranous disease of the throat, we shall obtain contradictory answers. Bretonneau answers, yes—a thousand times yes. Trosseau answers, yes; Guersant answers, yes. Indeed, there is a confidence in the power of these agents among the French physicians, whose opinions are best known to us, that is all but overwhelming. There is a persuasion in their eloquent praises and reiterated assurances that has forced me to doubt my own experience; and when disappointed in the effects of these agents, to return to them again and again in the hope that, by a closer imitation of their methods, I might participate in their triumphs.

"Among our own physicians I know some whose faith in the saving virtues of a timely and efficient application of these substances is not dimmed by a single doubt. I have a friend, judicious and observing, who can not convince himself that the throat membrane can ever resist the free application of solid nitrate of silver, or a solution of it, one hundred and twenty grains to the ounce of water, when it is used early and often. Dr. Woodward, of Brandon, Vermont, believes that he and his friend Dr. O'Dys owe a portion of their success (sixty cases without a single death) to the early use of this agent. This, and alterative doses of mercurials, were their main reliance; and he seems to suppose that if the disease, which was so fatal in a neighboring town, had been treated in the same way, the results would have been more favorable. On the other hand, while the English physicians generally are far behind the French in their praise of caustic applications, many, like Greenhow, object to them altogether, except in particular conditions. Greenhow's language is worth quoting: 'Local treatment applied to the throat internally has been almost universally adopted in the treatment of diptheria; and, though I by no means deny its value when judiciously employed, I am sure much mischief has been produced by its indiscriminate use, etc. * * * Observing that the removal of the exudation, and the application of remedies to the subjacent surface, neither shortened the duration nor sensibly modified the progress of the complaint, but that the false membrane rarely failed to be renewed in a few hours, I very soon discontinued this rough local medication.' When, however, the exudation is all within sight, and the surrounding parts are healthy, he thinks it proper to apply solid nitrate of silver, or nitric or muriatic acid, for he says: 'It is just possible in such cases that this treatment might check the progress of the complaint, and lead to a rapid recovery.' (Diptheria, pp. 263-4-5.) Dr. Turner, of Virginia, referring to similar applications (*American Medical Times*, Dec. 18, 1860), says: 'I studiously avoid probangs; I look upon them as instruments of torture and death. I know I have seen cases that died from the constant mopping to which the throat was subjected.' Dr. Metcalfe, of this city, says of the application of nitrate of silver to the throat (*American Medical Times*, Aug. 25, 1860), that he can not say he has derived any benefit from it. Indeed, in my intercourse with the physicians of this city, I meet but few who have not tried it, and disappointed in its promised benefits, have abandoned it. My own observation has taught me that the false membrane will not fall off by the mere application of nitrate of silver, either on the exudation or on the surrounding parts, without the use of some mechanical force; and that its application to tissues, after forced or spontaneous removal, will not prevent the reproduction of the exudation, at least in numbers of instances. I have seen the membrane appear when it was not looked for, in the course of scarlet fever for example, and where the nitrate of silver had been systematically applied for what appeared to be a different kind of sore throat. Yet, in these cases, it has sometimes followed upon the very heels of that medication. Such facts as these, however, do not *prove* that the application of the nitrate of silver is useless. They destroy our faith in its unfailing virtues, and fairly raise the

question, whether this kind of treatm
cruel, and to be abandoned; or if faili
really saves the lives of some. This d
solve for you. I can only say that n
curative powers of all caustic applica
shaken. But they are proper applicati
there is any ground left for faith in th
may know how to use them, not from
myself, but from one whose confidence
of silver, as the representative, and th
all, illuminates almost every page of a
I shall quote again from Bretonneau.
memoirs, he recommends hydrochlori
with three parts of honey; he even use
centrated and pure. Powdered alum
quent application. In his fifth memo
his former statements (p. 192 and on
local applications employed to modify
ulcerations, there are none so painful as
drochloric acid, while a solution of nitr
less painful and more efficacious;' an
credit of first suggesting it to Dr. Macl
gow. 'On the first day of the app
Egyptian chancre (meaning here tonsil
radical cure can be obtained in forty-ei
is sufficient to employ on the first day t
cations—one in the morning and one i
and to repeat the proceeding the ne
sponge used for the application shoul
not soaked. When the disease has pass
chea, the sponge should be applied wi
ure to the opening of the larynx, the
held pitilessly forward. 'After a few n
the same proceeding must be repeated i

the sponge having been washed, wiped, and dried, by pressure of a very dry piece of linen.' He relates the case of a child three years old, in which a membrane that was raised was a cast of the larynx, and its broken bronchial extremity had an alarming thickness, such as to forbid tracheotomy, but in which four applications in this way, each repeated (eight each day), were practiced. 'From the fourth day all anxiety ceased.' 'I affirm that without error in calculation, a solution of *thirty-two grammes* (four hundred and ninety-four grains) of the crystallized nitrate of silver was completely employed in this horrible treatment.' Two thirds at most being wasted; 'yet the rest was in great measure mingled with the mucous matter drawn in at the time of the cauterizations.' The linen washed and dried in the sun showed, by the black spots upon it, that unusual quantities of the salt had been swallowed.

"When the disease is detected in the nostrils, he advises to inject a solution of nitrate of silver with a padded syringe; and to inject both nostrils, especially if there is the least swelling of the neck glands on the two sides. Bretonneau does not inform us regarding the strength of the solution which he prefers, but the common practice is to make it forty to one hundred and twenty grains to the ounce of water."

Dr. Winne gives us the most promiscuous jumble of drug-medication extant, and as the best specimen of its kind I put it on record. If it does not convince the reader that the prevalent practice in diptheria is a series of blind experiments on the vitality of the patients, I know of no evidence that will be likely to do so.

"The local treatment consists chiefly in the application of caustic and astringent substances, in one form or another, to the affected part. Of these, the most usual are nitrate of silver, either solid or in solution, powdered alum, chloride of lime, chloride of soda, sesqui-chloride of iron, and hydrochloric acid.

"M. Bretonneau almost invariably employed the last of these remedies as a local application in his own practice, with the most marked success. The hydrochloric acid may be employed very nearly of the strength of the dilute acid of the shops, or considerably reduced in strength—dependent upon the severity or mildness of the attack. The best method of applying it is to moisten a small sponge attached to a probang or a camel's hair pencil with the fluid, and while depressing the tongue with the left hand, to carry the brush forward with the right, until the fauces are reached, when those parts of the tonsils, uvula, or soft palate on which the membranous deposit appears, may be moistened with the fluid, and the instrument withdrawn. The hydrochloric acid should be applied not only to the membranous surface, but to the parts immediately surrounding it, by which means the spread of the membrane is often arrested. The application should be renewed several times a day. Care, however, must be taken not to apply it of too great strength, or too often at the onset of the disease, especially if the symptoms are not of an aggravated character; otherwise the local disease may be enhanced, by the unnecessary injury inflicted upon the surrounding parts. The symptoms often appear momentarily aggravated by the local application, which is not unfrequently followed by an attempt to dislodge the membrane by vomiting. Should this latter result follow, the tonsils and palate will appear as if shrunken in substance, and spotted here and there with a few drops of blood upon the surface formerly occupied by the membrane.

"When this does occur, the application may be renewed directly upon the surface of the gland, in order to arrest the almost invariable disposition of the membrane to renew itself upon the abraded part. As the disease progresses, and the membrane extends toward or into the pharynx, the difficulty in making local applications becomes greatly enhanced; but the practitioner should not hesitate, for fear of inflicting temporary pain, from thoroughly exploring and covering the parts affected with the solution of hydrochloric acid. For the purpose of effecting this, it is often necessary to place the head of the patient upon the knee of an assistant, and with a spatula to depress the tongue and the lower jaw firmly at the same time, by which means a view of the whole fauces may be obtained, and an opportunity afforded of making a thorough application of the local remedy.

"Nitrate of silver has been warmly recommended by Trosseau, Guersant, and Valleix, in France, and was the application almost universally resorted to in England at the commencement of the epidemic in that country. The usual mode of using nitrate of silver in England was in solution. Dr. Kingsland advised a solution of 16 grains to an ounce of distilled water; and Dr. Hart, 30 grains to an ounce of distilled water. The mode of its use resembles that of the hydrochloric acid.

"When the local application of nitrate of silver is made in a solid form, care should be taken that it does not slip from the holder, or break, as in such an event it might fall into the stomach. Such an accident actually happened to M. Guersant; fortunately, however, the stomach rejected it; but this might not always occur, and few medical men would be willing to take so hazardous a risk. Dr. Hauner, of Austria, considers nitrate of silver as the very best local application to the diseased surface, and advises its use in a solution of from a scruple to half a dram, to an ounce of water.

"Subsequent experience did not confirm the good opinion entertained for nitrate of silver among the English practitioners, and many who were at first loud in its praises came to disuse it altogether. A substitute for this was found in the sesqui-chloride of iron, which is recommended by Dr. Ranking as being very efficacious in its effects upon the false membrane. He advises its use in the form of a gargle, of the strength of two drams to eight ounces of water, to be applied to the throat by means of a brush.

"In the United States, opinion appears to be divided as to the best local application. Dr. Blake, of Sacramento, has found the greatest benefit resulting from an application of strong hydrochloric acid; a view in which he is sustained by Dr. Bynum and Dr. Thomas, both of whom have had much experience in the treatment of the disease. Prof. Comegys, of Cincinnati, is in the habit of applying nitrate of silver, either in substance or strong solution in water. Sometimes, when the ulcerations are deep, he touches them with strong nitric acid, by means of a brush. In some cases he has employed with considerable benefit inhalations of tannic acid dissolved in sulphuric ether, applied by means of a cloth wetted with it, to the mouth. The formula is:

℞.—Tannic acid.......... .... f. ℨij.
Sulph. ether.. .... f. ℥j.   M.

"Dr. Jacobi, of New York, who, as physician to the Canal Street Dispensary, which treats a large number of German children, has had a very large experience, says:

"'The local treatment consists of cauterization of the membranes and surrounding parts with the solid nitrate of silver, or with strong or mild solutions of the same salt in water (ʒss-j. : ʒj.); of gargles, consisting of solutions of (or applying in substance) astringents, such as tannic acid, alum, sulphate of zinc, or claret wine; in gargling with, or applying, such medicinal agents as are known to have some effect on the constitution and tissue of the pseudo-membranes, as chloride of potassium, chlorates of potassa and soda, diluted or concentrated nitric or muriatic acids, liquor of sesqui-chloride of iron, etc. Astringents will prevent maceration, render the exudation dry and hard, and alter the consistency of the surrounding hyperæmic and edematous tissue. It will thus prevent, sometimes, the extension of pseudo-membranes to the neighborhood of the parts already affected, and in some cases may accelerate the expulsion of the membrane as a whole. We have thus seen the best effects from tannic acid, either applied directly to the parts by means of a curved whalebone probang, or dissolved in water as a gargle (ʒss-ii. : ʒi.) Of the tinct. sesquichlor. iron we have seen no particular effect. Cauterizations with nitrate of silver we have found to be generally of very little use when applied to the pharynx. Its effect is superficial only; it will form a scurf, but will destroy nothing. Destruction of the parts can not be effected except by forcing the caustic into and below the membrane; this can seldom be done in the pharynx of children, and for this reason cauterization is unavailing at this point, but will prove beneficial, we believe, by confining the process of exudation to its original locality. In cutaneous diptheria cauterization may be exercised to its full extent; but as these cases are generally attended with extreme prostration, the general treatment will prove both more necessary and successful. If cauterization is to be resorted to, we generally use, and with good effect, more or less concentrated muriatic, or acetic, or nitro-muriatic acid. Where, however, cauterizations are made, great caution is necessary not to mistake afterward the result of the caustic for pseudo-membrane. This remark is particularly applicable where nitrate of silver has been used.'

" Alum, chloride of lime, and calomel are sometimes recommended. When their use is deemed advisable, they may be applied by dipping a brush or the finger in the dry powder, and carrying it directly to the affected part, or blowing them through a quill.

" Prof. Metcalfe advises the use of the bromide of iodine, in the form of two drops to an ounce of the mucilage, or gum-arabic, as a topical application. He also gives dram doses of this mixture internally, with the happiest results.

" When there is a considerable accumulation in the nares and behind the velum, the *débris* and foul secretions may be removed, and much temporary relief obtained, by an injection of an infusion of chamomile with a few drops of creosote, which may be best effected by a laryngeal syringe. The syringe of Dr. Warren, of Boston, answers a very good purpose for injecting fluid either into the nares or below the epiglottis. It, however, is liable to the objection that it is likely to produce irritation, by coming in contact with the irritable portion, exactly at the opening of the glottis, which is found, by the researches of Prof. Horace Green, to be the seat of sensibility, instead of the epiglottis, as has heretofore been supposed. The common glass syringe, with either a curved extremity or a straight one—dependent upon the part to be reached—answers all ordinary purposes, and possesses the advantage of being easily obtained at the apothecary's, and is of slight cost.

"For correcting the fetor of the secretions, the chloride of soda, in the proportion of one dram to six ounces of water, may be used with much benefit. Dr. Ranking suggests, on the supposition of the presence of some vegetable parasite, the use of sulphurous acid and hyposulphate of soda, in the form of a saturated solution. 'The power of the latter,' he adds, 'in destroying the fungoid growth of favus, as well as the oidium which infests the vine, I have myself experienced; and I strongly recommend it, provided the vegetable origin of diptheria be confirmed by further observations.'

"Much relief is often afforded by inhalation, especially after the second or third day of the attack. An excellent means of fumigation is to pour boiling water upon catnip, or the leaves of any similar plant, with the addition of a little vinegar, and to allow the patient to inhale the fumes, either by inclosing the head under a blanket, or by applying the mouth to a tube connected with a close vessel containing the materials from which the vapor is generated. The immediate effect of fumigation is extremely grateful to the patient. Dr. Gurdon Buck advises the addition of Labarroque's solution of the chloride of soda, in successive portions of a teaspoonful each, to the liquid used for fumigation. Mr. C. T. Hodson recommends the inhalation of boiling water, to which has been added a tablespoonful of chlorinated lime.

"*General Treatment.*—The general treatment must be regulated by the type of the disease. Shortly after the appearance of M. Bretonneau's treatise, a great variety of treatment was recommended by different practitioners, all, however, with a view to arrest inflammatory action. Leeches to the neck, counter-irritation, especially by means of blisters, active mercurialization, and purgative medicines furnished the basis of most of the plans advised. Calomel, especially, obtained great celebrity, and was at one time considered as the most effective remedy in arresting the progress of the disease. It was first prescribed by Dr. Conolly, who was residing at Tours, at the appearance of the disease; and was so efficient in his hands, in minute doses, as speedily to find favor with the French practitioners. But, whatever may have been the success attendant upon its administration at that time, it is now found to require great caution in its use.

"Blisters are contra-indicated, and so far from furnishing relief, tend to increase the danger, by assuming an unhealthy, and frequently sloughy, appearance. The bites of leeches often give rise to passive bleeding, extremely difficult to arrest, which greatly reduces the already exhausted energies of the patient. Everything, in fact, which tends to lower the powers of life, or induce prostration, should be sedulously avoided, in the type of disease which at present prevails; and certainly differs from that for which Bretonneau, Conolly, and other medical men in France at that period were called upon to prescribe.

"The type of the disease as it now prevails exhibits a tendency to extreme prostration from the very beginning, and requires a tonic treatment to sustain the patient. The most effectual method of accomplishing this is by means of quinine, the various preparations of iron and steel, stimulants, in the form of brandy, milk punch, and wine whey, and a generous diet, consisting of beef-tea, Liebig's extract of meat, and a strong decoction of coffee. Sulphate of quinine may be administered in grain doses, conjoined to two grains of the sulphate of iron, repeated as often as the symptoms appear to require—usually every three hours. It is well to alternate this remedy with doses of chlorate of potassa, which appears to exercise a beneficial influence upon the disease of the mouth and throat. Chlorate of potassa may be given in doses of from five to ten grains, in distilled water, or a bitter infusion. Prof. Barker, of New York, advises the chlorate of potassa, in doses from ʒss. to ʒj. The chloride of soda has been recommended with the same intention, but does not appear to be equally efficacious with the chlorate of potassa.

"The tincture of the sesqui-chloride of iron has met with much favor among the English practitioners, as a tonic. Dr. Ranking gives it the preference to other tonics, although he frankly admits that it matters but little which of this class of medicines is used, provided the strength of the patient be sustained. 'Personally,' he remarks, 'I give the preference to the tincture of the sesqui-chloride of iron, not only from the inference drawn from the analogy of its unquestionable usefulness in the more asthenic forms of erysipelas, but also from the positive evidence of its benefit derived from the experience of several gentlemen in the country, among whom I may mention Mr. Dix, of Smallburg; Mr. Prentice, of North Walsham; and Mr. Cowles, of Stalham; each of which has had unusual opportunities of testing its advantages.' The tincture of the sesqui-chloride of iron may be administered in doses of from eight to sixteen drops, in a little water.

"Whatever may be the success or ultimate failure of this remedy, its first introduction into the treatment of this disease is undoubtedly due to Professor Thomas P. Heslop, of Queen's College, Birmingham, who, after repeated trials in his own practice, brought it to the attention of his clinical class at Queen's Hospital and the Medico-Chirurgical Society of Queen's College. His own success appears truly astonishing. 'I have given in this disease,' he says, 'to an adult twenty-five minims of the London tincture of the sesqui-chloride of iron every two, three, or four hours, and have conjoined a few drops of dilute hydrochloric acid. I have also applied daily, sometimes twice a day, by means of sponges, a solution of hydrochloric acid, but little weaker than the dilute acid of the London Pharmacopœia, and have always enjoined the regular use of weak gargles of the same acid. This, with the constant administration of stimulants, beef-tea, milk and jellies, has constituted my treatment; and I repeat here, what I have already stated in other quarters, that since I have become aware of the value of this medication, nearly ten months, I have not lost one case.' An excellent formula for administering a combination of chlorate of potassa and the sesqui-chloride of iron is: Chlorate of potassa, from eight to twenty grains; tincture sesqui-chloride of iron, ten to twenty-five drops; rose-water or orange-syrup, one dram; water, four ounces. Where there is difficulty in administering medicine, the bulk may be reduced by omitting the water altogether, and increasing at pleasure the amount of syrup. The success which has attended the use of this remedy in England warrants a careful trial of its merits at the hands of practitioners in the United States.

"Where the disturbance of the secretions appears to indicate the use of mercurial preparations, and they are not positively contra-indicated by the depressed state of the patient, calomel may be administered, in doses of one tenth of a grain, mixed with sugar, and placed dry upon the tongue. Dr. Bigelow has found this remedy valuable in the disease as it prevails at Paris; and Mr. Thompson was equally successful with it at Launceston, England. Dr. Anderson, of New York, and Dr. Briggs, of Richmond, have employed calomel with marked benefit. It is a question, when calomel and chlorate of potassa are administered conjointly, whether the effects of the potassa do not entirely annul those of the calomel. Dr. Bigelow, as the result of some very recent observations, says, that although it may retard or prevent the specific effects on the salivary glands, it does not in any way modify its action upon the secretions. It may be well, however, when the effect of the calomel is important, to intermit the use of chlorate of potassa for twenty-four hours, or to alternate the use of these medicines at wide intervals between the administration of the two.

"Emetics are serviceable when portions of the detached membrane are lodged in the throat, without being expelled, or when the disease is making rapid progress, and threatens to invade the larynx. The action of the emetic in this instance is frequently to detach the pellicle and dislodge the pseudo-membrane. At the same time that the membrane is thus ejected, the throat is relieved of the foul secretions which might otherwise be received into the stomach, to the great detriment of the patient.

"But, whatever treatment may be adopted, the fact should never be lost sight of, that the system is laboring under the influence of a powerful and most depressing poison; and it matters but little, so far as the constitutional treatment is concerned, whether this poison be at first local, and afterward disseminated through the system, or is from the beginning of a general character, and incidentally developed in the mucous membranes of the air-passages. In the performance of her functions in the elimination of this poison, Nature requires to be sustained, not only by the free use of the tonics already indicated, but by a liberal allowance of the most concentrated and nutritious articles of diet, in which beef-tea, milk, eggs, brandy, wine, and coffee stand prominent. When there is difficulty in swallowing, not only these articles of diet, but quinine, may be introduced, by means of injections; a resort to which should not be deferred until it is impossible to administer medicines by the mouth, but whenever the difficulty of swallowing becomes at all a prominent feature in the complaint. Injections should not be administered in greater quantities than two ounces at a time, and should not be often repeated; otherwise they will give rise to a local irritation in the rectum, which will prevent their retention. One or more drops of tincture of opii, according to the age of the patient, will greatly aid in the retention of the injection.

"After the violence of the disease has been checked, a continuance of the tonic treatment should be persevered in for some time, not only to prevent the sequelæ liable to follow, but a recurrence of the attack, which often reappears after an interval of several weeks, especially when the patient is exposed to those depressing influences which are too frequently attendant upon poverty and uncleanliness."

We have now seen what "confusion worse confounded" exists in the medical profession with regard to the treatment of diptheria, and how the testimony of medical men of equal character and experience is both *for* and *against* all plans of drug treatment which have yet been adopted. Some recommend the *stimulant* treatment; others prefer the *antiphlogistic;* some rely mainly on cauterizing the throat; others declare caustics to be injurious; some object to any strong *local applications* because the disease is *constitutional;* others object to powerful *constitutional* treatment because the disease is *local*, etc. And I will conclude this "budget of blunders" with a few quotations from the latest authors, showing that the discrepancies among physicians are still as wide and irreconcilable as ever. The especial object I have in view in dwelling so long on drug-medication is to destroy all confidence in it, and I know of no more effectual method of discrediting the system than that of telling the people what its advocates allege in its favor.

Dr. A. C. Hamlin, Surgeon to the Second Regiment Maine Volunteers, reports a case in which his treatment was chlorate of potash, gargles, iodine embrocations externally, inhalations of steam, and carbonate of ammonia and brandy, with *high diet*, internally; also cauterization with solid nitrate of silver; pieces of ice held in the mouth frequently; sponge baths, stimulants, etc. In detailing the plan and effects of treatment, Dr. Hamlin makes the significant statement, that after the cauterization the disease increased in both tonsils, and that, on applying the ice, there was an immediate improvement, a circumstance the importance of which we shall be better enabled to understand after we have examined the rationale and effects of Hygienic treatment.

Dr. Minot, Secretary of the Boston Society for Medical Improvement, has reported in the Boston *Medical and Surgical Journal* for March 21, 1861, the practice of several members of the Society : Dr. Lyman treated a case with chlorate of potash, fever mixture, solid nitrate of silver to the throat, castor-oil, wine, beef-tea.

Dr. Fifield stated that, in the cases he had seen, the application of solid caustic seemed to aggravate the disease, as did also the tincture of iodine.

Dr. Ainsworth prescribed strong solution of nitrate of silver, chlorate of potash, diluted muriatic acid, mustard to the throat and neck, citrate of magnesia, solid nitrate of silver, broth, flax-seed tea, strong solution of capsicum, per-chloride of iron, enema of strong beef-tea with Madeira wine, and wine by the mouth.

Dr. Minot reported a case in which " the treatment consisted in the administration of tonics, stimulants, and concentrated nourishment."

By "concentrated nourishment" the Doctor probably means *diluted* slops, broth, beef-tea and wine, brandy and toddy, of which we have already seen quite enough.

Dr. Jackson reported cases treated with quinine and muriate of iron internally, and muriatic acid to the throat. He states that nitrate of silver was at first applied to the throat, but seemed to do harm. An emetic and cathartic generally preceded the above treatment.

Dr. Tower, of South Weymouth, Mass., in a communication to Dr. Bowditch, published in the Boston *Medical and Surgical Journal* for March 7, 1861, says :

"The treatment which I have pursued has been various, but that which has found most favor with me is the free and frequent exhibition of chlorate of potassa; gargles of the same, or of water acidulated with muriatic acid, or, what is still better, a solution of common salt. The best external application is a saturated solution of common salt. I say this after trying various rubefacients and cataplasms. Cold water is employed by some, but I have never used it. If the disease is not arrested by these applications, I make use of a strong solution of nitrate of silver (ʒi. to ʒi.). When there is much prostration, stimulants, tonics, and plenty of beef-tea, or other nourishment."

Dr. W. A. Bryden, of Mayfield, in the *British Medical Journal* for Nov. 21, 1857, gives us his plan of treatment, which consists essentially in the use of guaiacum and chloride of potash, instead of the application of the solid nitrate of silver, which he regards as injurious.

Dr. Ramskid, of the Metropolitan Free Hospital, objects to strong caustic because, "in more than one case, it has seemed to increase every undesirable symptom." This plan of treatment is thus given in the London *Lancet* for Feb. 19, 1859, with a case to illustrate :

"A., a young lady, aged 15, of good condition in life, robust, and not subject to any of the influences supposed to be favorable to the development of the disease. I saw her on the third day, and found her in bed, tranquil, capable of speaking a few words together, breathing comfortably without noise, swallowing freely and without pain anything given to her; with a cool, soft skin, and silky pulse, beating 100. The submaxillary and cervical glands were much swollen. She could open her mouth tolerably well, and by means of a spoon a very distinct view of the interior was obtained. The soft palate was projecting, strongly convex on to the base of the tongue; the swelling eased off gradually, terminating on the hard palate, within half an inch of the front teeth; it was covered in patches with the characteristic false membrane, and everywhere exuded copiously a jelly-like, tenacious fibrin. By drawing forward and pressing down the tongue, the margin of the false palate could be distinctly seen, with its thick, swollen edge dipping down into the pharynx, and the uvula hanging in the center, pale-red and free from disease, or at most very slightly edematous. The tonsils were swollen, and agglutinated to the edges of the soft palate, and so matted with effusion of false membrane and fibrin as to be indistinguishable from the latter.

"According to the testimony present, the disease was decidedly progressing; there was more exudation, and the swelling was greater than six hours before. It was resolved to remove as much of the exudation as possible, no force being used, and to apply a strong solution of nitrate of silver, eighteen grains to the ounce. In four hours the breathing became noisy, not from implication of the larynx, but from blocking of the posterior nares, from increased swelling at the back part of the velum, and effusion of fibrinous secretion spotted with false membrane, and the corresponding difficulty of ingress of air by the mouth. Other measures, as inhalation and gargling, were adopted, and I may mention that the former always gave most relief. On the fourth day the report was, that the noisy breathing had considerably diminished; the patient had slept two hours and a half; at intervals of an hour, she had taken medicine, food, and wine, but with more difficulty than yesterday, and once the fluid returned by the nostrils; the anterior part of the palate was less swollen, and the false membrane and fibrin secreted in much less quantity; but the breathing was much more noisy than yesterday, and the cervical and submaxillary glands much more swollen; general symptoms, pulse, etc., as before. It was felt that the caustic application had done good in one direction and mischief in another; and the throat was mopped out, after removing gently all the exudation possible, with a solution of nitrate of silver ten grains to the ounce.

"For some hours the breathing was less noisy, but the difficulty recurred. Next day the report was—no sleep, increased difficulty in swallowing, and considerable accumulation about and behind the fauces; injection by the nares returned the same way; occasional smothered cough; the cervical, submaxillary, and neighboring glands immensely swollen. The mouth could not be opened wide enough to examine the throat. Laryngeal spasm occurred once, and was overcome by inhalation; again in two hours, and overcome by the same means. Sonorous respiration followed; and in an hour a third spasm of the glottis ushered in the fatal event.

"I know some persons may fail to see any connection between the application of the caustic and the increased swelling and aggravation of symptoms; and in this case it was contended by my consultant that the local treatment was the best possible under the circumstances, and that the fatal event would have occurred as soon under any other mode of treatment. I confess to having thought so at the time, but increased experience has convinced me that any treatment which causes rapidly increasing swelling of the cervical, submaxillary, and neighboring glands is bad, and sure to be attended by corresponding extension of the disease within and below the fauces, by declension of power, and increase in the difficulty of breathing, swallowing, etc. And the reason is obvious enough. In fact, in all cases where the first application of caustic has shown the tendency to excessive glandular enlargements, the local treatment can not be too soothing and gentle. The treatment consisted of quinine in three-grain doses, with ether and muriatic acid, and given in rotation with strong beef-tea and wine, so that every hour the patient took medicine, food, or wine.

"The treatment in which I have most faith, on or about the third day, under the circumstances above mentioned, is the following: Let as much of the exudation as is easily accessible and loose be removed. A carefully strained infusion of chamomiles is to be made, to which is added a few drops of creosote or of liquor calcis chlorinata (two drams to fifteen ounces), or liquor aluminis. It is to be used by means of Coxetter's laryngeal syringe, or any other apparatus the practitioner may advise, so as to avoid the unrest of tissues created by gargling. The laryngeal syringe is admirably adapted for children, who, after a time, will use it themselves, although, of course, not very effectually. If there be much accumulation behind the velum, and discharge passes by the nares, it is an exceedingly useful plan to syringe the throat through the anterior nares, with the same infusion. Most of the fluid comes back by the mouth, carrying with it *débris* of membrane and foul secretions. An effect at deglutition will almost always be made, but if some of the infusion be swallowed it can only do good. Inhalation from a hot infusion of the same has seemed to give more ease to the patient than any other application. Washing out of the throat should not be insisted on more than three or four times a day—the inhalation as often as the patient may wish. The chlorinated lime, in addition, should be used if there be much fetor; the alum when that is only slightly apparent. The throat outside should be surrounded by a poultice composed of the strained chamomile flowers, and changed four or five times in the twenty-four hours. Internally, I use quinine in chamomile infusion, with muriatic acid and ether, and endeavor to produce cinchonism. I use chamomiles, having in view its reputed efficacy in erysipelas and phlegmonous inflammation. In one case only, where hematuria existed, I gave tincture of acetate of iron, with acetate of potash in small quantity, and the patient did well."

Dr. Richard Cammack, in the *Lancet* for Oct. 30, 1858, gives us the following:

"*Treatment.*—1. A temperate, dry, well-ventilated room as can be obtained, no one being allowed to sleep in it except an attendant. Crowded bedrooms and animal effluvia are exciting causes.

"2. A calomel purgative, varying in strength according to the age and size of the patient; and in children, where symptoms of laryngitis appear; a rapid exhibition of the chloride of mercury, such as a grain to two grains every hour till the breathing is easier, and then every three or four hours, till the false membranes are loosened, the bowels evacuate green stools, or vomiting. Care is needed not to carry the mineral too far, but it can be borne in proportion to the strength of the patient and the sthenic form of the attack. Children who have been healthy, and are teething, have most inflammatory symptoms.

"3. The decoction of cinchona with hydrochloric acid, varying the dose of the latter from one minim to ten every four hours, in from a teaspoonful to two tablespoonfuls of the former.

"4. Gargle with chloride of sodium and vinegar, a tablespoonful of each in a teacupful of hot water; also inject this up the nostrils when they are becoming obstructed. This excels all other gargles; it relieves the breathing and the fetor, and causes the ulcers to heal.

"5. Apply the stick of nitrate of silver to every part where false membrane or exudation can be seen. By means of Dr. R. Quain's tongue depressor, one can see far and wide; but when the patient will not submit to this, and when the disease spreads beyond the reach of the caustic case, a probang and clean sponge well saturated with a strong solution of nitrate of silver will answer.

"6. Rub the external fauces with compound iodine ointment night and morning; and where erysipelas may appear, apply the stick, and lay on the plaster of strong mercury ointment.

"7. Keep the room and all else sweet and clean.

"8. A nutritious diet is necessary. A little mutton every day; boiled milk, rich gruels, and beef-tea, with hot port-wine and water (half wine with sugar and lemon), for all above ten years; and warm milk and water for minors. All things should be taken warm. Cold drinks are exciting causes.

"The disease is not infectious, except, perhaps, under extraordinary circumstances.

"Since I wrote the above remarks, I have seen many cases. I am convinced the malady is herpetic, and, therefore, would have called it so. It is malignant frequently, therefore herpes malignus anginosus would fully specify the disease. I have no wish to encroach on your space, but beg to observe that the medical gentlemen of this neighborhood are much at variance as to the nature of the disease, and that it has been very fatal. I have the *Lancet* from the commencement, and have read the lectures of most since Sir A. Cooper's and Mr. Abernethy's; but I have never seen a full description of this epidemic."

And next comes Dr. E. Peney, of Marden, Kent (*Medical Times and Gazette*, March 5, 1859), with turpentine as the leading remedial agent. Dr. Peney remarks:

"I have tried almost everything that I know to have been recommended, and have failed; and, perhaps, we often shall fail under any treatment; but I think it proper to mention a treatment which has been successful with me in three or four cases of late. It is for a child of from two to six years of age. Ten minims of the spiritus terebinthum every second hour, and five grains of the ammoniæ carbonæ every second hour, the child taking the turpentine one hour and the ammoniæ next hour.

"I rub up ʒij. of the spiritus terebinthum with the yolk of an egg, and add enough syrup to make a ʒxij. mixture. One teaspoonful in milk every two hours. Then dissolve ʒj. of the ammoniæ carbonæ in ʒxij. of water, and give one teaspoonful every two hours also in milk.

"Besides this the child takes port wine, porter, and beef-tea, or wine with the yolk of an egg *ad lib.* I have not found in any of my cases strangury caused by the turpentine. The patient dislikes it of course, and it requires a determined and attentive nurse; but I have found the plan very successful, and I speak of those cases where decided croupy breathing and fits of suffocation have made their appearance.

"I was induced to try the turpentine from having noted its effects, when given as advised by Mr. Carmichael in cases of iritis in broken-down constitutions where mercury could not be used, and where there is so great a tendency to the effusion of lymph in the chambers of the eye. We all know, too, how effectual it is in other diseases—acting like mercury in many respects—but stimulating instead of debilitating—and hence its appropriateness in diptheria, where mercury, I believe, hastens the fatal result. I now have recourse to no sponging the fauces with strong acid, or the argenti nitras, which I used to do, punishing a great deal and doing very little good."

Dr. J. C. S. Jennings, in the *British Medical Journal,* July 16, 1859, describes a plan of treatment which he claims to have been successful, and which is, in all important particulars, so far as the drug remedies are concerned, in direct antagonism with the stimulating plan of Professor Clark and others. Indeed, its leading agents are the most deadly antiphlogistics known to the materia medica. Dr. Jennings says:

"The plan I have invariably adopted, regardless of sex, or age, or incubation of disease, has been to give an emetic of antimonial wine, from half an ounce to an ounce, according to age; to freely cauterize the throat with solid nitrate of silver; to have a mustard poultice applied from ear to ear; the feet and legs plunged in a hot bath; and the patient confined to bed. After the emetic action has ceased, from three to five grains of calomel with five of compound extract of colocynth were given (or, for a child two grains of calomel with two grains of compound antimonial powder), and, four hours after, a mixture of bisulphate of quinia, chlorate of potash, and diluted hydrochloric acid. A gargle of chlorine solution was directed to be used frequently. When the inflammatory stage has been severe, the fauces tense and shining, and the throat edematous, spirit of nitrous ether and liquor of acetate of ammonia, or nitrate of potassa, has been added to the mixture.

"The diet has been at first farinaceous, and afterward consisting of strong broths and jellies. Stimulants have been very rarely administered, and then only as sherry whey, alternately with the quinine, which I have trusted to as the sheet-anchor. For infants, quinine may be given in jelly, washed down with a mixture of tincture of sesqui-chloride of iron. Too much stress can not be laid upon tartar emetic, quinine in large doses, and the *avoidance* or guarded use of alcoholic stimulants."

The treatment recommended and practiced by Dr. Smith, of St. Mary Cray, Kent (*British Medical Journal*, July 16, 1859), is essentially the opposite of that of Dr. Jennings; and as his experience and observations are, on several points, in direct conflict with those of several authors we have just quoted, I give his remarks in full:

"The principles that have guided my treatment of this disease are: *first*, to arrest the local inflammation by *exciting another* of a different character; *second*, to employ elimination according to the individual case; *third*, in all cases to sustain vigorously the vital powers.

"To accomplish the first indication, I prefer the employment of a strong solution of the nitrate of silver. Having first cleared the fauces, etc., as far as practicable by gentle means, I paint every affected part, and beyond it, with the solution, of the strength of fifteen grains to a dram. In mild cases I have frequently tried one of milder strength, say five grains; but I am satisfied that in all cases an efficient application of the full strength is the best. It is perfectly safe, and has at once a marked effect. It is more efficiently applied by a full-sized camel-hair pencil than a sponge. Severe cases must be seen again in twelve hours, and the application repeated should the so-called membrane spread. Later in the treatment, a weaker solution may be used, or Bretonneau's application, one part of hydrochloric acid to three of honey. And later still, when the membrane has disappeared, but much fullness and puffiness of the parts continue, a gargle, containing the sesqui-chloride of iron, or tannic acid. Where, as in my second case, there is much fetor, the chlorate of potassa is applicable. And where, as in my third case, there is more tonsillitis, we may, with advantage, employ inhalation of steam, or warm milk gargle. After the membrane is removed, and the tendency to diptheritic deposit supposed to be arrested, the throat must be carefully watched; for until the endemic condition of the system is conquered, we may have a relapse of diptheria.

"I commence the treatment of almost every case with a purge, varying with the state of the tongue, pulse, etc.; but by far the most frequently, calomel and rhubarb, carefully avoiding salines. In some cases, with loaded tongue and suffused countenances, I have given, with the greatest advantage, emetics. Indeed, I am now so satisfied of their value, that I shall for the future employ them more frequently, especially where the congestion is marked, or there is unusual tonsillitis. The further general treatment is of *great importance,* namely, that directed to sustain the vital powers and remove anæmia.

"I need not dwell upon the necessity of wine, beef-tea, etc. In the severe cases these are most urgently required, and must be liberally supplied. In the more trifling cases, if well marked, convalescence will be delayed, and danger of relapse continue, if these, or their equivalents, are not employed.

"Of all the medicines that may present themselves for our choice, there is one far superior, in my experience, to all others; and upon which I, indeed, chiefly rely: tincture of sesqui-chloride of iron. I have tried others that were obvious; but none sustain the vital powers, steady the pulse, lessen its frequency, and give potency to it; none remove the soft clam of the skin, steady the action of the kidney, and remove the anæmic pallor of the face, as does this. My confidence in its employment, and also in the use of the nitrate of silver, is fortified by their effects in erysipelas, in which they are almost specific. Cases will occur in which this treatment must be deferred, or modified, as where the tonsillitis is severe. In those cases, with the appropriate local treatment, I have first used the decoction of cinchona, with liquor of acetate of ammonia, or the latter with ammonia; but we afterward come to the steel.

"Such is a brief outline, and time admits of no

more, of the treatment of cases in which croup has not intervened. How are we to meet this formidable extension of the disease? Shall we, in any cases, resort to tracheotomy? I think not. Success, in reported cases, has not justified it; and we can not tell how far the membranous deposit has extended. I have had urgent cases of this description, and, happily, have hitherto treated them with success. My sheet-anchor is emetics, repeated, and very active ones, always of ipecacuanha and sulphate of zinc, never of antimony.

"Did time admit, I would detail these cases, but they present no peculiarity except the urgency of the symptoms. In one child, three years of age, I gave seven emetics before the symptoms were fully relieved. Portions of the membrane were detached and thrown off in the act of vomiting. I gave wine and ammonia in the intervals. In this case I gave also repeated small doses of calomel, because Bretonneau recommends it: and the case being of extreme urgency, I would not neglect one of such authority.

"In the more severe cases of diptheria, I can not too impressively recommend strict horizontal position. I have seen more than one case in which fatal syncope was to be apprehended if this had been neglected."

But a truce with druggery. We have had enough of it. We have been surfeited with the contradictory stories of their *virtues* and their bad effects, and with the absurd reasonings and conflicting statements of their advocates and authors. And I conclude this chapter of inconsistencies with an article written more than a century ago, and published in Boston in 1740. The discriminating reader will readily perceive that, however much physicians have progressed in the grammar school, they have, so far as the treatment of malignant diseases is concerned, "advanced backward" since the following article was published. It was written to a friend by a clergyman, in reference to what has since been called, "The Throat-Distemper of the Last Century," and which is supposed by many to be identical with the now prevalent diptheria.

"S<small>IR</small>—In Compliance with your Desire, I shall now communicate to you some of those Observations I have made upon that extraordinary Disease, which has made such awful Desolations in the Country, commonly called the Throat-Distemper.

"This Distemper first began in these Parts, in Febr. 1734,5. The long continuance and universal Spread of it among us, has given me abundant Opportunity to be acquainted with it in all its Forms.

"The first Assault was in a Family about ten Miles from me, which proved fatal to eight of the Children in about a Fortnight. Being called to visit the distressed Family, I found upon my arrival there, one of the Children newly dead, which gave me the Advantage of a Dissection, and thereby a better Acquaintance with the Nature of the Disease, than I could otherwise have had: From which (and other like) Observations, I came pretty early into the Methods of Cure that I have not yet seen Reason to change.

"There have few Distempers been ever known, that have put on a greater variety of Types, and appear'd with more different Symptoms, than this has done; which makes it necessary to be something particular in describing it, in order to set it in a just View, and to propose the Methods of Cure necessary in its several Appearances. And

"1. I take this Disease to be naturally an Eruptive

milliary Fever: and when it appears as such, it usually begins with a Shivering, a Chill, or with Stretching, or Yawning; which is quickly succeeded with a sore Throat, a Tumefaction of the Tonsils, Uvula and Epiglottis, and sometimes of the Jaws, and even of the whole Throat & Neck. The Fever is often acute, the Pulse quick & high, and the Countenance florid. The Tonsils first, and in a little Time the whole Throat covered with a whitish Crustula, the Tongue furr'd, and the Breath fetid. Upon the 2d, 3d, or 4th Day, if proper Methods are used, the Patient is covered with a milliary Eruption, in some exactly resembling the Measles, in others more like the Scarlet Fever (for which Distemper it has frequently been mistaken) but in others it very much resembles the confluent Small Pox. When the Eruption is finished, the Tumefaction everywhere subsides, the Fever abates, and the Slough in the Throat casts off and falls. The Eruption often disappears about the 6th or 7th Day; tho' it sometimes continues visible much longer. After the Eruption is over, the Cuticle scales and falls off, as in the Conclusion of Scarlet Fever. If after the Crise of this Disease Purging be neglected, the Sick may seem to recover Health and Strength for a while; yet they frequently in a little Time fall again into grievous Disorders; such as a great prostration of Strength, loss of Appetite, hectical Appearances, sometimes great Dimness of Sight, and often such a weakness in the Joints as deprives them of the Use of all their Limbs; and some of them are affected with scorbutick Symptoms of almost every Kind.

"When this Distemper appears in the Form now described, it is not very dangerous: I have seldom seen any die with it, unless by a sudden Looseness, that calls in the Eruptions; or by some very irregular Treatment. But there are several other very different Appearances of the Disease, which are attended with more frightful & deadly consequences.

"2. It frequently begins with a slight Indisposition, much resembling an ordinary Cold, with a listless habit, a slow & scarce discernible Fever, some soreness of the Throat and Tumefaction of the Tonsils; and perhaps a running of the Nose, the countenance pale, and the Eyes dull and heavy. The patient is not confin'd, nor any Danger apprehended for some Days, till the Fever gradually increases, the whole throat, and sometimes the Roof of the Mouth and Nostrils, are covered with a cankerous Crust, which corrodes the contiguous Parts, and frequently terminates in a mortal Gangreen, if not by seasonable Applications prevented. The Stomach is sometimes, and the Lungs often, covered with the same Crustula. The former Case is discovered by a vehement Sickness of the Stomach, a perpetual vomiting; and sometimes by ejecting of black or rusty and fetid Matter, having Scales like Bran mixed with it, which is a certain Index of a fatal Mortification.—When the Lungs are thus affected, the Patient is first afflicted·with a dry hollow Cough, which is quickly succeeded with an extraordinary Hoarseness and total Loss of the Voice, with the most distressing asthmatic Symptoms and difficulty of Breathing, under which the poor miserable Creature struggles, until released by a perfect Suffocation, or stoppage of Breath.—This last has been the fatal Symptom, under which the most have sunk, that have died in these Parts. And indeed there have comparatively but few recovered, whose Lungs have been thus affected. All that I have seen to get over this dreadful Symptom, have fallen into a Ptyalism or Salivation, equal to a petit Flux de Bouche, and have by their perpetual Cough expectorated incredible Quantities of a tough whitish Slough from their Lungs, for a considerable Time together. And on the other Hand, I have seen large Pieces of this Crust, several Inches long and near an Inch broad, torn from the Lungs by the vehemence of the Cough, without any Signs of Digestion, or possibility of obtaining it.

"Before I dismiss this Head, I must observe that the Fever which introduces the terrible Symptoms now described, does not always make such a slow and gradual Approach: but sometimes makes a fiercer Attack; and might probably be thrown off by the Eruptions, and this Train of Terrors prevented, if proper Methods were seasonably used.

"3. This Distemper sometimes appears in the Form of an Erysipelas. The Face suddenly inflames and swells, the Skin appears of a darkish Red, the Eyes are closed with the Tumefaction, which also sometimes extends through the whole Neck and Chest. Blisters or other small Ulcers here and there break out upon the Tumor, which corrode the adjacent Parts; and quickly bring on a Mortification, if not by some happy Means prevented. Some that are thus affected, are at the same time exercised with all the terrible internal Symptoms above described; and some with none of them. If this inflamed Tumor be not quickly discussed, it will (I think) always prove mortal.

"4. Another Appearance of this Disease is in external Ulcers; which break out frequently behind the Ears; sometimes they cover the whole Head and Forehead; sometimes they appear in the Arm-Pits, Groins, Navil, Bŭttocks or Seat; and sometimes in any of the extream Parts. These are covered with the same Kind of whitish Crustula above described, which also corrodes the contiguous Parts; and quickly, if not prevented, ends in a Mortification. I have ordinarily observed, that if these outward Ulcers are speedily cured, the Throat and internal Parts remain free from the above mentioned terrible Symptoms; otherwise the miserable Patient must pass thro' the whole tragical Scene of Terrors before represented, if an external Gangreen don't terminate his Agony and Life together.

"5. Sometimes this Disease appears first in Bubo's under the Ears, Jaws, or Chin, or in the Arm-Pits, or Groin. These, if quickly ripened, make a considerable Discharge; which brings a salutary end to the Disease; otherwise they quickly end in a fatal Mortification; or else bring on the whole foremention'd Tragedy.

"6. This disease appears sometimes in the Form of a Quinsey. The Lungs are inflamed, the Throat and especially the Epiglottis exceedingly tumefied. In a few Hours the Sick is brought to the Height of an Orthopnœa; and can not breathe but in an erect Posture, and then with great Difficulty and Noise. This may be distinguished from an Angina, by the Crustula in the Throat, which determines it to be a Sprout from the same Root with the Symptoms described above. In this Case the Patient sometimes dies in twenty-four Hours. I have not seen any one survive the third Day. But thro' the divine Goodness these symptoms have been more rarely seen among us, and there have been but few in this Manner snatched out of the world.

"As the Symptoms of this Distemper are very different, so the Methods of Cure should be respectively accommodated to them; and I shall therefore consider them distinctly.

"When this Distemper makes its Attack with the Symptoms of a high Fever, a florid Countenance &c. (as in the first Case described) the first Intention, to be pursued towards a Cure, is to bring out the Eruptions as soon as possible; to which End, I order the Patient to be confin'd in Bed, and put into a gentle breathing Sweat, till they appear. A Tea made with Virginian Snake-Root and English Saffron, with a few Grains of Cochineal; A Posset made with Carduus Mariæ boil'd in Milk, and turn'd with Wine, the Lapis contrayerva, or Gascoign-Powder; any or all of these, as occasions require, answer to this Purpose, and seldom fail of Success.

"One of the most dangerous Circumstances that attend this Disease, is a Looseness, that frequently happens upon the first Appearance of the Eruptions; which must be speedily restrain'd, and the Belly kept bound, lest the morbifick Matter evaporated by the Pores, be recalled into the Blood, and prove suddenly fatal.—To that Purpose, I ordinarily advise to Venice-Treacle, or liquid Laudanum, which commonly answer all intentions. But if the Patient should be in a dozing Habit, that these cannot be used, or if these should fail of Success, any other Astringent may be used that is proper in a Diarrhœa.

"The Ulcers in the Throat should be constantly cleansed, from the Time of their first Appearance. I have found the following Method most successful to this Purpose. Take Roman Vitriol, let it lie as near the fire as a Man can bear his Hand, till it be thoroughly calcined and turn'd white: Put about eight

Grains of this into half a Pint of Water; Lay down the Tongue with a Spatula; and gently wash off as much of the Crust as will easily separate, with a fine Ragg fastened to the End of a Probe or Stick, and wet in this liquor made warm. This Operation should be repeated every three or four Hours.

"After the Eruptions are quite gone, the Patient should be purged two or three Times, to prevent the Consequences above described; and this Rule should be observed in every Form of the Disease.

"If after the Crise of this Disease, in any of its Appearances, the Sick should fall into any of the Disorders mentioned under the first Head, such as Loss of Strength, a feverish Habit, Dimness of Sight, Weakness of the Joynts &c., Repeated Purging, as far as the Patient's Strength will bear, with Elixir Proprietatis, given twice a Day in a glass of generous Wine, will constantly remove these Difficulties.

"When this Disease makes a more slow and leisurely approach with a lingering Fever, pale Countenance &c. as described in the second Case, all Attempts to bring out the milliary Eruptions seem in vain. And therefore, tho' the Sick may be very much relieved by the diaphoretick Medicines abovementioned, if repeatedly used during the Course of the Illness; yet these are not to be depended upon for a Cure. But a brisk Purge should be also directed every third Day, and those Catharticks that are mixt with Calomel or Mercurius dulcis, are most likely to be serviceable, where the Age and Strength of the Patient will bear it.

"If there be an extream nauseating, and vehement Sickness of the Stomach, that can't be otherwise quieted, an Emetick seems necessary, tho' I have not found Encouragement to use vomiting Physick in any other Case.

"The internal Ulcers of the Throat should be treated as above directed; but if there be a great Tumefaction of the Glands, I order externally a Plaister of Diachylon cum Gummi and de Ranis cum Mercurio mixt; and internally the following Fumigation. Take Wormwood, Penny-royal, the Tops of St. John's Wort, Camomile-Flowers and Elder-Flowers, of each equal parts; boil very strong in Water; when boil'd, add as much Brandy or Rum as of this decoction; steam the Throat, thro' a Tunnel, as hot as can be born, three or four Times a Day.

"When the Lungs are seized with this cankerous Crustula, which is indicated by the Cough and Hoarseness above described, Mercurial Catharticks frequently repeated seem the best of any Thing to promote Expectoration. I have also found Success in the Use of the Syrup of red Poppies and Sperma Ceti mixt.

"When this Distemper appears in the Form of an Erysipelas, I have used the following Fomentation with good Success. Take Wormwood, Mint, Elder-Flowers, Camomile-Flowers, the Tops of St. John's. Wort, Fennel-Seeds pounded, and the lesser Centaury, equal Parts; Infuse in good Brandy or Jamaica Rum, in a Stone-Jugg well stop'd, and keep hot by the Fire: wet a Flannel Cloth with this; and after moderately squeezing out the Liquor, apply three or four double to the Tumor, as hot as can be born, every Hour.—In this Case I repeat Purging, as above directed.

"As for the external Ulcers above described (under the 4th Head) they may be always safely and speedily cured, by applying once or twice a Day a good thick Pledget of fine Tow dipt in the above described vitriolick water. I have never known this fail in a single Instance, when seasonably used. But then it must be observed, that some of these Ulcers will require this Water much sharper with the Vitriol, than others will bear. It should be so sharp as to bring off the Slough, dry up the flow of corrosive Humors, and promote a Digestion: but it must not be made a painful Caustick. In this the Practitioner's Discretion will guide him.

"I need not say any more respecting the Bubo's, mentioned under the fifth Head: but that they must by all possible Means be ripen'd as quick as they can; and launced as soon as they are digested and found to contain any Pus.

"I have not yet found any effectual Remedy in the 6th and last Case described.

"Upon the Disease in general, I have made the following Remarks; which perhaps may be of some Use.

"I have observ'd, that the more acute the Fever is on the first Seizure, the less dangerous; because there's more Hope of bringing out the Eruptions.

"I have observ'd, that there's more Danger of receiving Injury from a cold Air in this, than in any eruptive Fever I have seen. The Eruptions are easily struck in; and therefore there ought to be all possible Care, that the Sick be not at all exposed to the Air, till the Eruptions are quite over and gone.

"I have also observ'd, that there's much greater Danger from this disease in cold Weather, than in hot. In cold Weather it most commonly appears in the Form described under the second Head; while on the contrary, a hot Season very much forwards the Eruptions.

"I have frequently observ'd, that once having this

Disease is no Security against a second Attack. I have known the same person to have it four Times in one Year; the last of which prov'd mortal. I have known Numbers, that have passed thro' it in the eruptive Form in the Summer Season, that have died with it the succeeding Fall or Winter: tho' I have never seen any upon whom the Eruptions could be brought out more than once.

"I have ordinarily observ'd, that those who die with this Disease, have many Purple-Spots about them; which shews the Height of Malignity and Pestilential Quality in this terrible Distemper.

"Thus, Sir, I have endeavor'd in the most plain and familiar Manner to answer your Demands. I have not attempted a Philosophical Inquiry into the Nature of this Disease, nor a Rationale upon the Methods of Cure. I have meant no more than briefly to communicate to you some of my Experiences in this Distemper, which I presume is all you expect from me. If this proves of any Service, I shall have Cause of Thankfulness: If not, you'll kindly accept my willingness to serve you, and to contribute what I can towards the Relief of the afflicted and miserable. I am Sir,

"Your most humble Servant,
"Jonathan Dickinson.

"Elizabethtown, N. Jersey, *Febr.* 20, 1738,9.

"postscript.

"Since I wrote this Letter, I am inform'd by a Gentleman of the Profession, who has had very great Improvement in this Distemper, That he has found out a Method of Cure, which seldom fails of Success in all the Forms of this Disease herein described, (the first, fourth, and fifth only excepted, which should be treated as above directed) and that is a Decoction of the Root of the Dart Weed, or (as it is here called) the Squaw Root. He orders about an Ounce of this Root to be boiled in a Quart of Water, to which he adds when strain'd a Jill of Rum and two Ounces of Loaf-Sugar; and boils again to the consumption of one quarter Part. This he gives his Patients frequently to drink, and with this orders them frequently to gargle their Throats; allowing no internal Medicine but this only, during the whole Course of the Disease, excepting a Purge or two in the Conclusion. I have seen a surprising Effect of this Method in one Instance; and shall make what further Observations I can: And if this answers my present Hopes, I shall endeavor to give you further Information.

"The Dart-Weed grows with a strait stalk six or eight Foot high, is jointed every eight or ten Inches apart; and bears a large white Tassell on the Top, when in the Flower. The Root is black and bitterish."

HYGIENIC TREATMENT OF DIPTHERIA.

Having seen what merit there is in the drug treatment of diptheria, and what reliance can be placed on the theories and experience of medical men, who believe in a system which is in opposition to Nature, contrary to common sense, and in direct antagonism with every law of the vital organism, let us now proceed to consider the rational treatment of the disease.

As I have already explained, diptheria consists essentially of a local inflammation and a general fever. In many cases the throat affection, which is the local inflammation, is slight, while the constitutional affection, or general fever, is severe; and in other cases the reverse happens—the local affection being severe and the fever slight. The fever is always of the low, atonic, and typhoid character. The local inflammation, in all severe cases, is attended with an excretion of coagulable lymph, which, concreting into a false membrane, forms a preternatural crust or coating to the mucous surface, to be cast off, like all other foreign or abnormal substances. When spread over a large portion of the larynx, trachea, or bronchial ramifications, this membranous concretion may occasion death by suffocation. In the other cases which terminate fatally, death is the result of exhaustion.

The *cause* of diptheria is poison, virus, or impurities of some kind in the blood. The *disease itself* is an effort of the system to purify itself by expelling these impurities. When the remedial effort is chiefly directed to the surface, there will be much constitutional disturbance of the kind denominated *fever*. If the determination to the surface is attended with considerable heat and dryness of the skin, it may be mistaken for *high* or *sthenic* fever. When the process of purification is determined chiefly to the mucous membrane, there will be corresponding disturbance of the function of the part, and of the character which medical authors recognize as *inflammation*. When the whole mass of blood is very gross, and the determination to the throat very violent, ulceration and disorganization of the structure follow rapidly, and the disease takes the name of "putrid sore throat." In many cases the process of depuration is very nearly equally divided between the skin and mucous membrane, in which case the life of the patient usually depends on the kind of medication—whether it increases the determination to or from the external surface. Medicines may be employed which do not materially unbalance nor derange the existing remedial effort, and although they are really a damage to the patient, and prolong the convalescence, yet because they are not appreciably mischievous at the moment, they may get the credit of curing the disease.

When the remedial struggle is nearly balanced, or directed chiefly to the external surface, there is very little danger, and such cases seldom terminate fatally, except as death is the result of maltreatment. The danger results from the concentration of morbid action to a particular point, thus disorganizing and destroying the tissue; hence the danger may be measured, as a general rule, by the violence of the throat-affection. There are cases, however, in which the system is so gross, the blood so impure, and all the fluids so foul, that before the remedial effort has become established in the direction of any outlet, the patient will sink and die of exhaustion, with very slight manifestations of general fever or of local inflammation.

When the patient is blessed with a good constitution, and his habits of living and exposure to infection are not such as to render his blood and secretions greatly depraved, the remedial effort—the process of purification—will be so equally balanced and so well maintained, that he will bear a great amount of injurious treatment, and endure a hundred doses of drug poisons, without losing his life. But if, on the contrary, the constitution is very frail or very gross, so that the morbid action is directed wholly *from* the surface, a small bleeding, a single leech, a blister, a cathartic dose, a mercurial purge, or an antimonial emetic, or a single touch of the burning caustic, may decide the case against the patient in a few hours.

The danger of diptheria results, usually, from the

excessive determination of morbid action to the throat; and, hence, the obvious indication of cure is to counteract this determination by promoting depuration in other directions, especially through the skin. By counteracting this determination of morbid action, or of remedial effort—for, however strange the language may seem to persons unaccustomed to it, these phrases really mean the same thing—I do not mean *repressing* or *subduing* it, but *regulating* it.

And here is the great principle which underlies all correct medication, and which forms the broad distinction between Hygienic and Drug Treatment. I do not look upon disease as a thing to be "subdued," "suppressed," "destroyed," "expelled," or exterminated. *It is an action to be regulated.* To regulate remedial effort, or morbid action, is simply so to control and direct it that each organ or part may perform its own appropriate duty, to the end that no structure may be disorganized by having too great a burden thrown upon it. Instead of subduing disease by merely opposing or counteracting the symptoms, the proper business of the physician is so to diffuse, direct, and equalize it, that it may successfully accomplish its work of purification.

The first indication, then, in the treatment of diptheria, is *to balance the circulation*, and in fulfilling this indication the temperature of the body is the proper and the infallible guide. Wherever there is deficient circulation there are coldness and paleness, and wherever there is congestion or obstruction there are pain, heat, and disturbed function; and these conditions must ever be kept in mind, as they are the basis of all proper therapeutic applications and processes.

We have seen that the disease may be attended with

all grades and shades of *typhoid* fever, and with all degrees of *atonic* local inflammation. The constitutional disturbance which we denominate fever, may be attended with much or little preternatural heat of the surface, or with none at all; or with a temperature below the normal standard; or with irregular temperature—some parts of the surface being above and others below the normal standard.

The general remedial plan, therefore, so far as the fever is concerned, is resolved into the simple idea of *regulating the temperature.* To do this is to promote equal distribution of the blood, in other words, to balance the circulation; and as functional action is always in proportion to the vigor of the circulation, so by regulating the temperature and balancing the circulation, we supply the conditions which enable "Nature" to cure the disease; or, in less figurative phraseology, which aid and assist the living system to do its work of purification.

The cure of disease consists in removing the causes; not in silencing the remedial struggle; for this is but killing the patient.

If these views are correct—and no medical man will ever seriously controvert them—it is easy to understand how it is, and why it is, that a bleeding, or a blister, or a dose of Epsom salts, or an antiphlogistic operation of niter, antimony, colchicum, digitalis, aconite, or veratria may quickly extinguish the patient's lamp of life, by concentrating remedial action in the center of the vital domain, when all of the living energies are needed to determine the morbid matter to the surface.

They show, too, the great delusion of the medical profession, in relying on stimulants or antiphlogistics to promote action to or from the surface. All that these agents do, or can do, is to occasion or aggravate a fever, or induce or aggravate a local inflammation, thus adding to the causes of disease, and necessitating a waste of vital power to get rid of them.

Nature does not own, and the living system abhors this whole plan—though it be the plan of the whole medical profession—of " curing a primary disease by creating a drug disease."

To balance the circulation, regulate the temperature, promote external depuration, and remove congestion and obstruction, we do not need the inflaming stimulants, the corroding caustics, the paralyzing narcotics, nor the deadly antiphlogistics. We have in water alone all that is usable or useful for the purpose indicated. It may be employed of any temperature from ice to steam, according to the circumstances of any given case. Water is the sole vehicle by means of which all of the nutrient materials of the body are transported to the various structures, by which all of the effete materials or waste matters of the body are carried to the various outlets, and is also the material by which the temperature of the system is properly radiated, balanced, maintained, and regulated. There is nothing provided in the universe that can subserve these purposes except water. And if it plays so important a part in the normal exercise of the functions, it becomes even more necessary, if possible, in their abnormal exercise—the state of disease—when there is extra and unusual duty to perform.

In the state of health, and under all the ordinary circumstances of life, the temperature of the body is easily regulated, and the circulation balanced, so that disease is *prevented*, by means of air, exercise, clothing, and artificial heat. In health the external use of water is often refreshing and invigorating, and conducive to longevity; but in disease it becomes a necessity. In disease there are venoms, viruses, poisons, or accumulated impurities of some kind, to be diluted and washed away; and this calls for a more free use of water internally than is demanded in the state of health. And as the solvent and detergent properties of water are in the direct ratio of its purity, how absurd is the practice of medicating the water—whether it is to be employed externally to regulate temperature, or internally to cleanse the system of noxious matters—with mustard, vinegar, saleratus, salt, ashes, spirits, roots, herbs, barks, leaves, flowers, seeds! etc. Although these foreign substances do not in all cases prevent the water from having some degree of beneficial effect, they always diminish its value in proportion to their quantity. There would be just as much sense, reason, or science in taking the impure water of the ocean to cook victuals or wash clothes with, as to employ water holding in solution mineral, earthy, or alkaline ingredients to cleanse the solids and purify the fluids of the living system.

It is true that there are persons who call themselves "Hydropathic" physicians, and who are the proprietors of what they advertise as "Water-Cure" establishments, who recommend and prescribe "mineral waters;" but I could never understand why the poisons or impurities taken from "medicinal springs" are so different from the *same* poisons or impurities obtained at the apothecary shop.

Before proceeding to explain the proper treatment for diptheria, I will, in order to save repetition, copy from a small work I have published—"Water-Cure for the

Million," a description of the various bathing processes, so far as they may be applicable to home-treatment:

"1. WET-SHEET PACKING.—On a bed or mattress two or three comfortables or bed-quilts are spread; over them a pair of flannel blankets; and lastly, a wet sheet (rather coarse linen is best) wrung out lightly. The patient, undressed, lies down flat on the back, and is quickly enveloped in the sheet, blanket, and other bedding. The head must be well raised with pillows, and care must be taken to have the feet well wrapped. If the feet do not warm with the rest of the body, a jug of hot water should be applied; and if there is tendency to headache, several folds of a cold wet cloth should be laid over the forehead. The usual time for remaining in the pack is from forty to sixty minutes. It may be followed by the plunge, half-bath, rubbing wet sheet, or towel-wash, according to circumstances. The pack is not intended as a sweating process, as many suppose, though a moderate perspiration is not objectionable. A comfortable temperature of the surface is the desideratum, independent of more or less sweating, or none at all. When the patient warms up rapidly, thirty minutes or less will be long enough to remain enveloped; but when he becomes warm slowly and with difficulty, an hour, or more, is not too long. In some cases it is necessary to put hot bottles to the sides as well as to the feet. When the object is to cool a fever, the sheet should be allowed to retain more water, or if the skin is very hot, double sheets may be used. In chronic diseases, when the main object is to induce 'reaction,' or rather circulation, toward the surface, the sheet should be wrung more thoroughly, and the patient enveloped with a greater quantity of blankets, comfortables, or other bedding.

"2. HALF-PACK.—This is the same as the preceding, with the exception that the neck and extremities are not covered by the wet sheet, which is applied merely to the trunk of the body, from the armpits to the hips. It is adapted to those whose circulation is too feeble for a full pack; it is also often employed as a preparation for the full pack.

"3. HALF-BATH.—An oval or oblong tub is most convenient, though any vessel allowing a patient to sit down with the legs extended will answer. The water should cover the lower extremities and about half of the abdomen. While in the bath, the patient, if able, should rub the lower extremities, while the attendant rubs the chest, back, and abdomen.

"4. HIP OR SITZ-BATH.—Any small-sized wash-tub will do for this, although tubs constructed with a straight back, and raised four or five inches from the floor, are much the most agreeable. The water should just cover the hips and lower part of the abdomen. A blanket should be thrown over the patient, who will find it also useful to rub or knead the abdomen with the hand or fingers during the bath.

"5. FOOT-BATH.—Any small vessel, as a pail, will answer. Usually the water should be about ankle-deep; but very delicate invalids, or extremely susceptible persons should not have the water more than half an inch to one inch in depth. During the bath, the feet should be kept in gentle motion. Walking foot-baths are excellent in warm weather, where a cool stream can be found.

"6. WET AND COLD FOOT-BATH.—Place the feet in water as warm as can be borne for five to ten minutes; then dip them for a moment in cold water, and wipe dry.

"7. Rubbing Wet-Sheet.—If the sheet is used *drippingly* wet, the patient stands in the tub; if wrung so as not to drip, it may be used on a carpet or in any place. The sheet is thrown around the body, which it completely envelops below the neck; the attendant rubs the body over the sheet (not with it), the patient exercising himself at the same time by rubbing in front.

"8. Pail-Douche.—This means simply pouring water over the chest and shoulders from a pail.

"9. Stream-Douche.—A stream of water may be applied to the part or parts affected, by pouring from a pitcher or other convenient vessel, held as high as possible; or a barrel or keg may be elevated for the purpose, having a tub of any desired size. The power will be proportional to the amount of water in the reservoir.

"10. Towel or Sponge Bath.—Rubbing the whole surface with a coarse wet towel or sponge, followed by a dry sheet or towels, constitutes this process.

"11. Affusion Bath.—This implies pouring water gently over the surface of the body. The patient may stand in a tub, or lie on the bed, the bedding being protected by a sheet of India-rubber or gutta-percha.

"12. The Plunge-Bath.—This is employed but little, except at the establishments. Those who have conveniences will often find it one of the best processes. Any tub or box holding water enough to allow the whole body to be immersed, with the limbs extended, answers the purpose. A very good plunge can be made of a large cask cut in two near the middle. It is a useful precaution to wet the head before taking this bath.

"13. Drop-Bath.—A vessel, filled with *very cold* water, is furnished with a small aperture through which the water falls in drops. It is adapted to torpid muscles, paralytic limbs, tumors, etc. It should be followed by active friction.

"14. THE SWEATING-PACK.—To produce perspiration, the patient is packed in the flannel blanket and other bedding, as mentioned in No. 1, omitting the wet sheet. Some persons will perspire in less than an hour; others require several hours. This is the severest of the Water-Cure processes, and, in fact, is very seldom called for. The warm, hot, or vapor-baths are, in most cases, preferable.

"15. HEAD-BATH.—The patient lies extended on a rug or mattress, the head resting in a shallow basin or bowl, holding two or three inches of water, the shoulders being supported by a pillow. It is principally employed in chronic affections of the head, eyes, and ears. Wet cloths applied to the head, the "pouring-bath," and the "wet cap" are good substitutes.

"16. THE POURING HEAD-BATH.—The patient lies face downward, the head supported by an attendant, projecting over the side of the bed, which is protected by a sheet or blanket thrown around the patient's neck; a tub is placed under the head to catch the water, which is poured from a pitcher moderately, but steadily, for several minutes, or until the head is well cooled, the stream being principally applied to the temples and back part of the head. It is useful in severe cases of sick headache; in the early stage of violent choleras; in the early stages of fevers, when attended with great gastric irritation or biliary disturbance. In hysteria, apoplexy, delirium-tremens, nose-bleeding, inflammation of the brain, ophthalmia, otitis, etc., it has been employed with advantage.

"17. FOUNTAIN, OR SPRAY-BATH.—This consists of a

number of small streams of water directed to a particular part of the body. It may be regarded as a gentle douche or local shower. It is intended to excite action and promote absorption in the part or organ to which it is applied.

"18. THE SHOWER-BATH.—This needs no description. It is not frequently used in Water-Cure, but is often very convenient. Those liable to a "rush of blood to the head," should not allow much of the shock of the stream upon the head. Feeble persons, should never use this bath until prepared by other treatment. Placing the feet for a few minutes in warm water, before taking the shower, is a good preparatory measure for feeble persons. Standing in warm water, ankle deep, will materially lessen its shock on the brain and nervous system.

"19. NASAL, MOUTH, AND EYE BATHS.—Drawing water gently up the nostrils and ejecting it by the mouth, holding water in the mouth, and holding the eyes open in water of a temperature suited to the case, are the processes indicated by these terms. They are useful in relaxed and inflammatory affections of the mucous membranes and other structures of the parts.

"20. ARM AND LEG BATHS.—The limbs may be held in any convenient vessel containing the requisite depth of water. These baths are useful in cases of fever sores, chronic ulcers, inflammatory affections of the joints, etc.

"21. VAPOR-BATH.—Hot stones or bricks may be employed to generate vapor or steam. The patient may sit naked on an open-work chair, with blankets pinned around the neck; a small tub or a common tin pan, holding a quart of water, is placed under the chair, and red-hot bricks or stones occasionally put in the vessel, so as to keep the vapor constantly rising from the surface of the water. Another very simple plan is this: Procure a one-gallon tin boiler, with a half-inch tin pipe, having two or three joints and a single elbow. The boiler may be heated on any ordinary stove, grate, or furnace, and the pipe so attached to it as to convey the steam under the chair in which the patient sits, covered from the neck downward with blankets. It may be employed from ten to thirty minutes, according to the amount of vapor generated.

" 22. AIR-BATH.—The whole body is suddenly exposed to cool or cold air, or even to a strong current, and an excellent and invigorating process it is in many cases. There is no danger from it, provided the surface has a comfortable glow or temperature at the time, and the circulation is maintained by active exercise. Friction with the hand, a sheet, towel, or flesh-brush, is beneficial at the same time.

" 23. BANDAGES AND COMPRESSES.—These are wet cloths, applied to any weak, sore, hot, painful, or diseased part, and renewed so often as they become dry or very warm. The best surgeons have, in all ages, employed 'water-dressings' alone in local wounds, injuries, and inflammations. They may be *warming* or *cooling* to the part, as they are covered, or not, with dry cloths.

" 24. THE WET-GIRDLE.—Three or four yards of crash toweling make a good one. One half of it is wet and applied around the abdomen, followed by the dry half to cover it. It should be wetted so often as it becomes dry. It is extensively employed in bilious and dyspeptic affections, female weaknesses, etc. When required to be worn for a long time, it should, after the first few weeks, be omitted occasionally, or worn only a part of each day, so that the skin over which it is applied will not become too tender. It should not be worn when it occasions permanent chilliness.

"25. THE CHEST-WRAPPER.—This is made of coarse linen, to fit the trunk like an under-shirt, from the neck to the lower ribs; it is applied so wet as possible without dripping, and covered by a similar dry wrapper, made of Canton or light woolen flannel. It requires renewing two or three times a day. It is useful in most cases of pneumonia, asthma, consumption, bronchitis, etc. The same precautions apply to its prolonged employment as mentioned under the head of the wet-girdle.

"26. FOMENTATIONS.—These are employed for relaxing muscles, relieving spasms, griping, nervous headache, etc. Any cloths wet in hot water and applied so warm as can be borne, generally answer the purpose; but flannel cloths dipped in hot water, and wrung nearly dry in another cloth or handkerchief, so as to steam the part moderately, are the most efficient sedatives. They are usually employed from five to fifteen minutes. They are useful in cases of severe constipations, colic, dysmenorrhea, hysteria, etc.

"27. REFRIGERATION.—One part of common salt to two parts of snow or pounded ice makes a good freezing mixture. It is inclosed in a very thin cloth, and applied for a few minutes, until the requisite degree of congelation has taken place. It is useful in felons, styes, malignant tumors and ulcers, fever sores, cancers, and in some forms of neuralgia and rheumatism.

"28. WET DRESS BATH.—This is a method of self-packing, enabling the patient to dispense with the services of an attendant. A linen sheet is fashioned into the form of a night-dress, with large sleeves, and after the bed is prepared, the dress can be wet and put on; the patient can then get into bed and wrap himself sufficiently to secure a comfortable reaction.

"29. Electro-Chemical Bath.— A copper-lined bath-tub is necessary for this process. The patient is immersed in warm water up to the neck; one hand is brought in contact with the positive pole of a strong galvanic battery, the negative pole being in contact with the metallic lining of the tub. The water is usually acidulated, though in some cases alkalies are employed. From half a pint to a pint of nitric acid is put into the water for each bath. It should not be mixed with the water until the galvanic circuit is completed, either by having the patient in connection with the poles of the battery, or these in contact with the copper-lining of the bath-tub. The patient may remain in the bath from ten minutes to half an hour. This bath is very useful in a torpid condition of the skin with low circulation; in glandular obstructions; scrofulas, rheumatic and gouty affections; in chronic congestions of the liver, and to aid the elimination of mineral medicines and other poisons.

"30. Injections.—These are warm or tepid, cool or cold. The former are used to quiet pain and produce free discharges; the latter to check excessive evacuations and strengthen the bowels. For the former purpose so large a quantity should be used as the bowels can conveniently receive; and for the latter purpose only a small quantity—so much as can be conveniently retained. Small enemas of very cold water are highly serviceable in cases of piles, prolapsus, fissures, etc. The self-injecting syringe is the most convenient instrument. With a rectal, vaginal, and intra-uterine tube, it will answer all possible purposes, for old or young, male or female. These articles can all be furnished for $3.

"31. GENERAL BATHING RULES.—Never bathe soon after eating. The most powerful baths should be taken when the stomach is most empty. No full bath should be taken less than three hours after a full meal. Great heat or profuse perspiration are no objections to going into cold water, provided the respiration is not disturbed, and the patient is not greatly fatigued or exhausted. The body should always be comfortably warm at the time of taking any cold bath. Exercise, friction, dry-wrapping, or fire may be resorted to, according to circumstances. Very feeble persons should commence treatment with warm or tepid water, gradually lowering the temperature. All shocks, such as shower-baths, douches, plunges, etc., should be avoided by every feeble and irritable invalid; by consumptives in the second and later stages; by those who are liable to great local determinations, or congestions, as "rush of blood to the head," bleeding from the stomach or lungs, etc.; in displacements of the bowels or uterus; during the menstrual period of females; during any considerable crisis or critical effort; after the crisis or "turn" of any fever, or other acute disease; during the existence of any powerful emotion or excitement; soon after eating or copious drinking; in all cases attended with profuse discharges, as diarrhea, cholera, diabetes, hemorrhages; during the suppurative stage of extensive abscesses or ulcers. The heat or feverishness which may attend any of the conditions or diseases above-named should always be abated by tepid effusions or spongings. It is dangerous to employ the wet-sheet pack, in prolonged or violent fevers, after the crisis or turn of the fever. Many errors have been committed in ignorance of this rule. Never eat immediately after bathing.

"32. DURATION OF BATHS.—Many errors are committed by remaining in cold baths for too long a time. I have known cases in which dyspeptics and consumptives, at Water-Cure establishments, were kept in cold sitz-baths for two hours at a time, once or twice a day. This was intended as a derivative measure, but it worked very injuriously for the patients. Derivative baths, like all others, must be determined by the condition of the patient, not by the thermometer nor chronometer. Sitz-baths of a mild temperature should seldom be prolonged beyond twenty minutes; more frequently ten to fifteen minutes are preferable. It is better to repeat all bathing appliances frequently, than to make violent impressions less frequently. Plunges, douches, and showers, if the water is cold or cool, should not ordinarily be continued more than a minute; when the temperature of the water is temperate, or tepid, they may be taken from five to ten minutes. Tepid half-baths should usually be taken from five to ten minutes. Sitz-baths, foot-baths, head-baths, arm and leg baths, etc., may vary from five to thirty minutes. But, as already intimated, regard must always be had to the temperature of the water and the circulation of the patient."

In the premonitory stage of diptheria, when the patient is affected with rigors or chilliness, or these with alternate and irregular flushes of heat, a full warm bath, as warm as the patient can comfortably bear, for ten or fifteen minutes, should be employed if practicable. If this is impracticable, the warm hip-bath and the hot foot-bath are the best substitutes. If these are not available, warm fomentations to the abdomen, and bottles of hot water to the sides and feet, should be resorted to.

If there is at this time pain or soreness of the throat without much heat, fomentations should be applied externally for ten or fifteen minutes, followed by the cold wet compress covered with a dry towel or cloth; and this should be re-wet and re-applied as often as it becomes warm or nearly dry. The patient should keep entirely quiet at this time, avoiding everything in the shape of food, condiments, stimulants, and medicines, swallowing nothing but pure water, and of this only so much as is demanded by the thirst. The temperature of the water for drinking may be that which is most agreeable to the patient.

When the hot stage of the fever is fully developed, the tepid half-bath, the tepid ablution, or the wet-sheet pack may be resorted to. The wet-sheet pack is best adapted to those cases in which the heat and dryness are uniform over the whole surface, and the patient is not greatly prostrated. But with more feeble patients, and when the external heat is moderate or unequal, the tepid ablution should be preferred. The tepid half-bath is applicable to the same cases as the wet-sheet, and is only preferable when the attendants do not well understand the management of the packing process. The temperature of the water should be cool, but not very cold—from 75° to 85° Fahrenheit. Either of the baths may be repeated as often as occasion requires; that is, as often as the external temperature of the patient rises much above the normal standard. The patient should be put to bed immediately after each bath and kept warm and comfortable; much sweating, however, is to be avoided, although a very moderate perspiration may be desirable. Too much care can not be taken to keep the feet warm and the head cool; and if there is the least tendency to coldness of the lower extremities and heat of the head, hot bottles should be applied to the feet and cold wet cloths to the head.

When there is much pain, heat, or swelling of the throat, or when the little patches of fibrinous exudation become visible on the mucous membrane of the tonsils, or elsewhere, the local treatment must be varied accordingly. Cold applications must be now resorted to, and employed thoroughly and perseveringly until the morbid secretion is arrested. Sips of very cold water may be taken frequently, or, what is still better, bits of ice may be put into the mouth and allowed to melt away, while the throat is enveloped in cold wet cloths. The indication now is to check the violence of the inflammation, and thus arrest the exudation of the membranous material; and there is no way to do this so certainly and so effectually as to reduce the morbid heat below the point which is essential to the excretion of coagulable lymph. This plan has been employed in thousands of cases of croup with almost uniform success. And in the ulcerated sore throat of malignant scarlet fever, it is the only reliable resource.

If the patient is at any time troubled with harassing cough, difficult expectoration, or laborious respiration, or all together, after the violence of the inflammation has subsided, warm-water-drinking, to the extent of inducing vomiting if need be, should be resorted to. And in the later stage of the disease, when the concreted exudation is firmly adherent to the mucous surface, or has extended widely in the bronchial tubes, warm fomentations to the chest and throat are proper.

To these cases the moist atmosphere, or vapor, as recommended by Dr. Sayre, is especially adapted.

Although the fever may in some cases be violent, so far as severity of symptoms is concerned, and the throat affection intense, so far as the inflammatory action is concerned, yet as the diathesis is always atonic or asthenic, these conditions will much sooner yield to the proper cooling remedies named than they will in the truly entonic or sthenic diathesis. In what is properly denominated entonic visceral inflammation, that is, enteric or sthenic fever, with active or phlegmonous inflammation, patients will frequently bear to be packed in double wet-sheets, with advantage; and these may frequently be repeated two or three times in twenty-four hours. But the preternatural heat is never so persistent in diptheria. The wet-sheet pack rarely requires more than one application in twenty-four hours, and it seldom happens that more than two or three repetitions of this process are necessary to materially mitigate the violence of the febrile action; after which, should the skin incline to feverishness, the tepid ablution will be sufficient.

Abundance of *pure fresh air* is quite as important in the treatment of diptheria as are the bathing processes. No doubt the contagiousness or non-contagiousness of the disease depends very much on the means which are employed to ventilate and cleanse the apartment. The patient should be kept comfortable, by means of bed-clothes, and fire if necessary, but on no consideration should all of the windows and doors be closed for a moment. The safety of others as well as of the patient may depend on this precaution. In close rooms, and in underground apartments, where free ventilation by doors and windows is impossible, the air of the place may be purified to a great extent by swinging the door vigorously forward and backward. In this way, in the crowded tenement houses of our cities and large villages, fresh air may be supplied and the accumulated miasms expelled, when there is no other possible method for "raising the breeze."

The purifying and invigorating influence of *light* and *sunshine* should never be disregarded. They are useful in nearly all morbid conditions, and of especial value in the management of putrid and infectious febrile and inflammatory diseases. When practicable, the rays of the sun should be admitted freely into the sick chamber, and, during the day, the room should be as well lighted as possible. But at night, when the external senses of the patient need quiet and repose, light should be excluded; nor should the talking or whispering of watchers be allowed in the room.

Rest is an important element in the Hygienic treatment of disease. And there is no remedial resource so little understood and so generally disregarded as this. A majority of physicians seem to have no idea of its necessity or value; or if they do, they ignore it altogether in practice. Indeed, rest is out of the question if the patient must be disturbed and disquieted with some dose, or drug, or slop, every hour or every half hour, and even awakened out of sleep, if he is so fortunate as to be able to slumber, to swallow something which does more harm than good.

The chief point of skill in the true physician is to know when to let the patient alone. It is easy to perceive morbid phenomena, and to combat symptoms; but to know when Nature is doing just right, and when she should not be interfered with, require judgment and discrimination. "Let-alone-ativeness" is the chief merit of Homeopathy. The patient is amused with infinitesimal placebos, while Nature has time and opportunity to remove the causes of disease, and then the cure results as the necessary consequence of having nothing more for the *vis medicatrix* to do.

So far as food and diet are concerned, very little need be said. As I have already intimated, the practice of stuffing the patient continually on what the doctors, with consistent absurdity, call "nourishing diet," is exceedingly pernicious. During the acute stage of the disease, while the fever is violent and the inflammation severe, no food can be digested, and none should be taken. As the fever subsides, the patient may be allowed a little gruel, and good fruit, to be followed, as convalescence advances, with such farinaceous articles as mealy potatoes, beans, peas, unleavened bread, etc. Baked apples, tomatoes, stewed or raw, sweet oranges, etc., may generally be allowed as freely as the patient desires, and, until the crisis of the disease is fairly passed, no other food is required.

Drink may be taken according to thirst; but when there is great thirst with a disposition to vomit, very small draughts should be taken and frequently repeated. There is no objection in such cases to the juices of acid and subacid fruits, properly diluted, as lemon juice, apple water, oranges, etc. Dried berries, stewed and slightly sweetened, will answer, in some cases, for both victuals and drink. In New York city, and probably in most other parts of the country, dried blackberries, raspberries, and whortleberries can be had in abundance in the winter season. Preserved berries can also be found, nearly as fresh and savory as when first picked from the bushes. The value of a really *frugivorous* diet, in febrile diseases, has never been sufficiently appreciated.

Enemas may be necessary in the early stage of the disease, but are seldom required afterward. In the outset of the disease, provided there has been no diarrhea, the bowels should be moved freely with an injection of tepid water; and subsequently only when a sense of fullness and distention of the abdomen indicates the presence of accumulated fecal matters.

When *vomiting* becomes a troublesome symptom or complication, small bits of ice may be occasionally swallowed, or frequent sips of cold water taken, and a cold wet towel covered with a dry cloth should be applied over the region of the stomach.

Diarrhea can be relieved by means of warm fomentations to the abdomen, and small enemas of cool or cold water, administered immediately after the evacuations. The patient should keep the horizontal posture, and be as quiet as possible.

Albuminaria does not require any special medication.

Hemorrhages can generally be promptly checked with cold applications.

Extreme swelling of the glands of the neck requires the constant application of wet cloths to the part.

Coma may be relieved by cold applications to the head and warm ones to the feet. In extreme cases, fomentations to the abdomen may be employed advantageously.

The *sequelæ* of diptheria demand only a strict attention to the general health, except so far as they are the effects of drug-medicines, and then all the appliances for purification must be brought into requisition.

That the plan of treatment I have now detailed is successful, I have not only my own experience, and that of other physicians of the Hygeio-Therapeutic School, to offer as evidence, but I have also the testimony of some of the drug doctors themselves. As an illustration, I will give, in full, an article published in the Dansville (N. Y.) *Herald:*

"DIPTHERIA, SORE THROAT, AND QUINSY SUCCESSFULLY TREATED BY THE LOCAL APPLICATION OF ICE.

"DANSVILLE, *Feb.* 18, 1861.

"MR. EDITOR—*Dear Sir:* Allow me, through the columns of your valuable journal, to make some practical remarks on the subject of diptheria. I shall confine myself to the consideration of its early symptoms, and its early or abortive treatment. I do not propose to enter into a lengthy discussion as to what is or what is not diptheria, except so far as to make myself understood as to the treatment, that being what I most desire to bring to the notice of your patrons and the public generally. Hence, sir, what I have to say will be as strictly practical as may be. If, by any course of treatment, the early or premonitory symptoms can be stopped, then we have no diptheria.

"Now, sir, I have had this disease to treat constantly for the last twenty months, and what I have to say is the result of actual observation and experience; therefore it is not the result of mere speculation and theory, of which we have had quite enough. I conceive it to be about time for somebody to bring the subject down to facts, and these facts sustained by a uniform success in practice, applied according to the rules which experience has found necessary to be observed.

"Then, sir, we lay it down as an axiom, that diptheria in its early stages is nothing more nor less than an inflammation, and that there never was a case of diptheria without a preceding inflammation to a greater or less extent, and that the inflammation has a termination peculiar to itself.

"Again we lay it down as an axiom, that no man, however close his observation, can distinguish between an inflammation, the termination of which will be diptheria, and one the termination of which will be pus, as in common tonsillitis, quinsy, or gangrene and sloughing, as in putrid sore throat. While the inflammation is being developed previous to its termination, the result can not be foretold with any degree of certainty. That inflammation of the throat results in one of the three mentioned forms, daily experience verifies; that is to say, in an exudation which immediately becomes organized tissue, forming the false membrane, which constitutes a case of diptheria. This membrane may be in small and separate patches, or it may extend all over the back part of the mouth and upper part of the windpipe, and even, as it sometimes does, travel down into the smaller bronchial tubes. It is the formation of this false membrane which constitutes a case of diptheria.

"Then as to the other two terminations of inflammation of the throat (and one that is scarcely less fatal), we may say that one is that of mortification or sloughing, called putrid sore throat, the other is in the formation of matter or pus; this is designated quinsy or tonsillitis.

"Having thus briefly stated what experience bears me out in saying in relation to the early stages of this truly frightful disease, before stating the treatment proper, I wish to say a word in relation to what I think an erroneous and unsafe theory—that is, that the disease is *constitutional*, and that the soreness of the throat is but the local manifestation of a general or constitutional diseased action. Now, sir, I hold the reverse to be the truth ; that is, that the disease of the throat *is the disease,* and that the constitution becomes affected by absorption of the poison from the throat, the same as in the case of vaccination, when the mere speck of matter inserted under the skin of the arm produces a general affection, viz., kine-pox. I hold this theory of the primary taint of symptoms to be unsafe, from the fact that it misleads the physician, and his poor patient has been caused to swallow large doses of drastic purges and the like, in order, as he says, to rid the poor victim of some imaginary poison. But above all it has caused him to neglect the *proper* treatment of the throat trouble, and thereby allowing the only chance to slip ; and this very poison which is so much dreaded, time to absorb into the system at large, and this to bring on a fatal typhoid train of symptoms.

"Then to recapitulate. Diptheria, in its early stages, is but an inflammation, having a termination peculiar to itself, yet subject to the same laws that govern other inflammations, viz., heat, redness, and swelling, producing soreness just in proportion to the amount of inflammation, and the fever which attends is in exact ratio to the amount of local or throat trouble. Believing that I have made myself capable of being understood, I will now proceed with the treatment.

"First, then, envelop the neck in cloths wrung out of cold water (it is not the water but the cold), changing them as often as they get warm. If there is much swelling near the angle of the jaw, apply a bladder with a handful of *snow,* so arranged as to form a small surface.· This should be placed directly over the tonsils. So much for the external applications.

"If there is not much swelling externally, the cold cloths or snow may be omitted, and the case may be trusted to internal applications. The patient should go to bed, and laying upon the back should take a small piece of *ice* into the mouth and allow it to settle as far down as possible without swallowing it. When this has melted, he should spit out the water and have a fresh piece of *ice* applied. This will require a faithful and attentive nurse. The pieces of ice should be about as large as the first joint of the finger. By pursuing this course in the early stages of the disease, it will be cured in from *twelve to twenty-four hours.* If the disease has got a little further advanced, yet in its inflammatory stages, with high febrile action, put the patient into a warm bath, keep him there until he feels faint; take him out, wrap him in warm flannel blankets and sweat him for one or two hours, after which maintain a gentle perspiration for three or four days; for be it known that absolute rest for this length of time is essential. The ice is to be continued at the same time.

"This may seem to be a very simple treatment for so formidable a malady; but that is a mistake, for we have not a more powerful remedial agent than *ice* when properly applied; besides being formidable it is capable of perfect management, and all that is necessary is to *graduate the amount of cold to the degree of inflammation.* It also has another valuable feature, that of always being on hand, especially at this time of year.

"I am confident, from a long experience in the use of this remedy, that if people will observe and apply as above directed, that diptheria will be shorn of its terror and many a valuable life saved. Remember that it must be used early, from the first accession of throat symptoms, and persevered in until they are removed.

"This treatment may and probably will meet with the same reception that all great principles have when first brought before the world of mind—that is, that it is an innovation, and some of the wise old ones will shake their heads doubtingly. But I hold that disease when it can be cured should be, whether it be according to authority or without authority. Sir, where would be the mighty improvements that have been made in medicine, if nobody should take a step beyond authority? Respectfully yours,

"Z. H. BLAKE, M.D."

I am decidedly opposed to sweating any diptheritic patient for one or two hours. Many patients will bear it, and all may if the sweating be not too profuse; nor would I make it a point to keep the patient in the warm bath until he feels faint. A warm bath of ten or fifteen minutes' duration is sufficient, and if all faintness is avoided so much the better. It is well, afterward, to keep the patient quiet, and the skin in a moist, perspirable state; but anything like profuse sweating is to be deprecated.

Dr. Blake makes no allusion to "Water-Cure," or "Hydropathy," nor does he give the least hint that he ever knew or heard of a case of diptheria being treated with cold water and colder ice, with external warm or cold bathing, and without drugs of any kind, except in his own practice. So far as one can infer from his article, this practice with him is entirely original. The statement, "I have had this disease to treat constantly for twenty months, and what I have to say is the result of actual observation and experience," may be interpreted in various ways. His observations may have been made on the cases which were treated by other physicians, and who have treated it in the way he recommends.

At all events, it is true that many cases of diptheria were treated in Dansville, and several of them at the water-cure of Dr. Jackson, in that place, and all successfully, previous to the date of Dr. Blake's article. And it is also true that the *Water-Cure Journal*, which circulates largely in Dansville, had previously and repeatedly advocated a similar plan of treatment.

The theory advanced by Dr. Blake, that the causes of the disease are essentially local, and that the constitutional disturbance results from the absorption of the local infection, has been sufficiently refuted in the preceding part of this work. But as Dr. Blake predicates on the theory which he adopts a very plausible argument against the employment of drug-remedies, it seems needless to correct the error, so far as diptheria is concerned. But the principle involved applies to other diseases—indeed, to all diseases.

Dr. Blake objects to the internal use of poisonous drugs, because the causes of the disease are not in the blood, but on the mucous membrane of the throat. Will not this reasoning apply to other diseases as well as diptheria? Again, if drug-medication is proper *per se*, and if diptheria is primarily a mere throat affection, why not apply drugs to the throat? If drugs are really and properly curative agents, here is one of the best imaginable opportunities for employing them judiciously and successfully, because we can *see* the diseased part, and have the evidence of our senses as to the *modus operandi* and effects of the medicines. Here is inflammation, for which bleeding, niter, antimony, digitalis, salts, veratria, arnica, aconite, gelseminum, and all the host of antiphlogistics and narcotics have such a reputation for curing; and here is (according to Dr. Blake) a locally generated virus, and what mortal doctor of the drug school ever conceived the possibility of arresting, correcting, suppressing, destroying, or killing, or curing a virus without a specific drug, or a counter-poison, or an " alterative," without mercury in some form?

But, no. Dr. Blake proposes simply to *cool* the virus, to *refrigerate* the inflammation. He relies on *temperature* alone to destroy the infection, arrest the inflammation, prevent or remove the fever, and restore the patient to health. He is right in practice, but wrong in theory. Should he adopt the true theory and give the correct explanation, he could not long maintain before the world the position of drug doctor.

Dr. Blake reasons that, because the disease, or its cause, is local, it can be cured without drugs. But, admitting the disease, or its cause, to exist in the blood, why can not it also then be removed or cured without drugs? Dr. Blake's *practice* is revolutionary; and I am unwilling that an ingenious sophistication shall be allowed to save the theory of drug-medication from its damaging influence.

In all the places which I have visited during the last year, I have made special inquiries as to the prevalence of diptheria, the manner in which it has been treated by the physicians, and the rate of mortality. And all the information I have been able to collect from others, agrees precisely with my own observations and experience. Wherever the Hygienic plan of treatment, substantially as recommended in this work, has been adopted at the commencement of the disease, and persevered in to the end, to the total exclusion of all drug-medication, local or constitutional, no death has yet come to my knowledge. In three or four instances, where the patients were badly scrofulous, or very gross in dietetic habits, and where the physician was not called until the membranous exudation had extended to the bronchial tubes, the cases have terminated fatally. But such can hardly be regarded as exceptions to the uniform success of Hygienic treatment.

I do not claim, nor do I believe, that all cases are curable by the means which I recommend. No doubt there are cases which are incurable by any means whatever. There are, undoubtedly, persons so gross in body, so depraved in blood, so frail in organization, or so feeble in vital resources, that the existence of diptheria necessitates death. But these cases are exceptions, and rare ones too, to the general rule.

I have heard from more than one dozen of the graduates of the New York Hygeio-Therapeutic College, who have treated each rom one to twenty cases of diptheria and putrid sore throat, without as yet losing a single patient.

While riding on the cars from Iowa City to Chicago, in the month of January last, I made the acquaintance of Mr. C. Manfull, of Augusta, Ohio, who was returning from a trip to the West. Mr. Manfull informed me that, seven or eight years ago, he purchased the "Hydropathic Encyclopedia," and subscribed for the *Water-Cure Journal*, since which time he has treated his own and many of his neighbors' children, when sick of croup, diptheria, or any form of sore throat, with invariable success, having never lost a patient. He informed me also that an eminent allopathic physician in Steubenville, Ohio, where the diptheria had been extensively prevalent and very fatal, had possessed himself of the "Encyclopedia," adopted the Hygienic treatment, and abandoned all drug-medication, after which very few deaths occurred in the place. He stated, moreover, that Dr. Beaumont, of Cumberland, Va., had treated many cases of croup, diptheria, and malignant scarlet fever hygienically, and had not lost one patient. Dr. Beaumont has delivered public lectures on "Hygienic *versus* Drug-Medication," with good effect.

TRACHEOTOMY.

As a last resort, when the false membrane is so obstructing the air-passages as to endanger immediate suffocation, this operation is recommended by some authors. It is at least highly probable that some lives have been saved by the operation, but there is reason, too, to believe that, in some cases, life has been destroyed by it. It is not always possible to determine, whether the patient survives the operation or not, what influence the measure had in determining the result.

During the year 1856 there were fifty-four operations of tracheotomy for croup, at the Children's Hospital in Paris. Of these cases fifteen recovered. M. Guersant testifies that, in the cases in which he has operated, about one third have recovered. M. Bouchat operated on one hundred and sixty, and only five were saved. M. Bretonneau performed the operation in twenty cases, of which six recovered. M. Velpeau operated ten times, and two of his patients recovered. M. Perit operated in six cases, and in three of the cases the patients were saved.

The results of three hundred and eighty operations,

reported by M. Chaillon, were, two hundred and ninety-four deaths, and eighty-six recoveries.

The statistics of the Hospital des Enfants show a mortality of five to one.

In Great Britain, so far as the statistics have been reported, the results of the operation have been somewhat less favorable than in France.

The statistics of American authors are exceedingly meager on this subject, but do not vary materially from the reports of the French and British hospitals.

Dr. Gross, of Louisville, Kentucky, has published the particulars of one hundred and seventy-six cases of foreign bodies in the larynx; in sixty-eight of these cases the operation of tracheotomy was performed, with a mortality of only eleven per cent. But the success or propriety of the operation, in these cases, must be predicated on very different premises from those which apply to the necessity or utility of the operation in diptheria.

It is true that the operation of tracheotomy is not in itself a very difficult nor dangerous operation in adults, yet with young children the case is very different, and requires the utmost surgical skill and dexterity. And when the patient is extremely exhausted, the pain and alarm necessarily attending the operation might be sufficient to turn the scale against the patient.

Many authors have objected to the operation on the ground that it is apt to induce severe bronchitis, or greatly to aggravate the previously existing inflammation.

The only condition in which the operation can be called for or justified is when the diptheritic exudation has extended to the larnyx, and has become so firmly concreted and adherent to the membrane as to threaten death by suffocation. In these cases it is obviously possible to keep up the respiration by means of an artificial opening into the windpipe, until the false membrane can be cast off and expelled. It is, of course, a desperate expedient, and so much so that many practitioners of eminence and experience proscribe it entirely. There can be no doubt that the operation has, in many cases, probably in a great majority, been resorted to when the patient was actually moribund, so that the deaths were scarcely at all influenced by it.

The proper time for performing the operation, provided it be proper in any case, is not very precisely determined by the authors who have written on the subject. "We should not wait until the case is desperate, or the patient in a dying condition," says one; nor, says another, "should we attempt the operation too early, before other remedies had been fairly and completely tested."

But as to what precise time may be regarded as the proper "middle period," and how we are to know when all other remedies have been "fairly and completely tested," we are left entirely in the dark. Dr. Slade quotes approvingly the following rule as to time: "so soon as ever we feel that our remedies are too tardy to overtake the disease."

This may be an excellent rule for the *conscience*, but a very poor one for the *judgment*. A physician may practice very conscientiously, yet very injudiciously. The important information which the authors do not give us is, by what symptoms are we to know when to perform tracheotomy?

There is, I apprehend, no better rule to be governed by in determining this question than the one I have already intimated. When the patient is in a state of actual suffocation from the presence of the false membrane in the larynx, and the strength not greatly exhausted, the operation will be justifiable; but whether it will even then increase or decrease the chance of recovery, is a problem which I regard as by no means to be settled by the data before us.

The operation consists in making an opening into the windpipe, a short distance below the larynx, and introducing a canula, through which respiration can go on. It is important that the canula be large enough, or suffocation would soon take place; and great care must be taken to keep the instrument free, or respiration may cease from a stoppage of the tube. The general custom is to allow the tube to remain four or five days, and renew it should difficult breathing recur on its removal.

Tubing of the glottis, an experiment introduced by M. Bouchat, has been resorted to by other practitioners. The process consists in inserting into the larynx, through the mouth, a metallic tube, through which respiration is to be maintained. The most that is pretended in favor of this operation is, that it may delay asphyxia, and perhaps postpone for awhile the necessity for tracheotomy.

STIMULATION VS. ANTIPHLOGISTICATION.

I have now placed before the reader all the important facts and theories I can find in medical books and journals concerning the nature and treatment of diptheria, with the opinions of medical writers and teachers for or against the various methods of medication which have been proposed; and an explanation of the Hygienic plan of treatment, with the reasons therefor.

But I can not conclude this work satisfactorily to myself without a chapter devoted especially to the refutation of the gross error of the medical profession, and the great delusion of the people, not only as respects the nature and treatment of diptheria, but with regard to the proper management of all diseases.

Medical men always act from some recognition of a theory, however vague and indefinite it may be. Whatever doctrine or hypothesis the physician entertains respecting the intrinsic nature of any malady, it will in some manner influence his prescriptions at the bedside of the patient.

It is true that a large class of practitioners, finding by experience that all the doctrines of medical books and schools are unsatisfactory, that all of the practice recommended by the standard authorities is uncertain, and learning, too, by repeated disappointments, that the principles which medical authors teach will seldom apply in practice, have ignored all theory, and profess to be guided only by facts. Their only guides in the treatment of disease are their own observations and the experience of their predecessors. But as it happens that the observations of medical men are widely different, and the experience of their predecessors (being interpreted so as to agree with whatever theories they happen to entertain) is as contradictory as is possible to be, these guides seem to be extremely fallacious.

Medical authors have been contending for several centuries whether the stimulating plan of treatment, or just the opposite, the antiphlogistic, is the proper one for the treatment of certain febrile and inflammatory diseases, and thus are quite as far from any common agreement now as they were three hundred years ago. Is it not strange that not one of them has ever thought of the *primary question* which underlies this discussion—*is either method right?*

It is taken for granted that if a disease, or the patient, will not bear *depletion*, he must have *repletion*. If he can not endure *antiphlogistics*, he must be dosed with *stimulants*. If he sinks under *reducing* treatment, he must be "supported" with fiery irritants. If he will not tolerate bleeding, he must be fed with brandy; and if he can not digest wholesome food, he must be stuffed and gorged on such medico-dietetic abominations as alcoholized animal broths, grog-and-chicken tea, "strong nourishment" of wine-and-soup, etc., etc.

And, on the other hand, if stimulation seems to damage the patient, the antiphlogistic plan must of necessity do good. If the patient can not bear "supporting" treatment, he must have the opposite—the reducing. If brandy disagrees, bleeding must be in order, etc.

The whole error lies in assuming what is not true. Both practices are wrong. The indication of treatment is to *purify*, not to stimulate nor antiphlogisticate. The majority of physicians of the drug school recommend bleeding and reducing measures in "inflammatory states" of the system, and alcoholic and other stimulants in "typhous conditions." But what are inflammatory states, and what are typhous conditions? Here all is confusion again. As we have seen, some authors regard the fever of diptheria as inflammatory or sthenic, while others regard it as atonic or typhoid. And the same disagreement exists as to the diathesis of the throat affection.

The idea that stimulants "support the system," "impart energy," or temporarily "augment vitality," is the cause of nearly all the malpractice among medical men, and of all the dissipation and debauchery in the world. Stimulants *exhaust* vitality, as do antiphlogistics. Brandy and bleeding, opium and niter, quinine and antimony, rum and digitalis, capsicum and veratria, alike occasion the expenditure and waste of vital power, as do all poisons of whatever name or nature; and the notion that a poison which is intrinsically inimical to anything that has organic life, can support vitality in any degree or in any sense, is one of the wildest vagaries that ever possessed the minds of human beings.

The grand mistake of medical men on this subject arises from a false theory of the *modus operandi* of medicines. It is everywhere taught in medical books and schools, that *medicines act* on the different parts and organs of the body in virtue of their "inherent affinities" for those organs. Nothing can be more absurd. The truth is exactly the contrary. The *living system acts on the medicines*. It acts on poisons to expel them from the vital domain. Some it expels through the skin by a prompt vigorous determination of blood and nervous energy to the cutaneous emunctory; this process is attended with a feverish state of the system and increased heat of the surface; this abnormal excitement or fever is called "stimulation;" and the article or agent which occasions it is said to be a "stimulant."

The *effect* of the medicine, or the poison, is a fever or an inflammation, and nothing else. And a fever or an inflammation can not "impart vitality" to the system. Nothing can impart what it does not possess. It can not "support" the machinery of life. It is the same precisely whether the fever or the inflammation—

the disease—be occasioned by medicine, poison, indigestible food, " catching cold," a wound, an injury of any kind, or any other cause. If it exists at all, it is abnormal action; and abnormal action always *expends* and never augments vital power.

Antiphlogistics also occasion the waste and loss of vital power, but in a different direction, and hence the morbid phenomena are very different. They divert action *from* the surface; in other words, they are resisted by a determination of blood and nervous energy from the circumference of the body to the center, thus occasioning symptoms the very opposite of those which are called stimulation. The skin is cooler, and the pulse weaker, and the muscular power, instead of being preternaturally excited, is directly depressed. So far as the *effects* of stimulants and antiphlogistics are concerned, stimulants may be said to be *indirectly*, and antiphlogistics *directly*, exhausting.

A similar controversy has long existed in the medical profession respecting the theory of inflammation. By some authors inflammation is regarded as an *increased action* of the blood-vessels of the part, or, as some teachers express it, " inflammation is an augmentation of all the vital powers of that part which is the seat of it ;" while others contend that it is just the contrary, a *decreased action* of the blood-vessels, or a diminution of all the vital powers of that part which is the seat of it. And this controversy is apparently no nearer a settlement now than it was a thousand years ago.

The subject, however, seems to be important; for if the theory of increased action be true, the antiphlogistic plan of treatment seems to be indicated; while if the doctrine of diminished action be correct, the stimulating plan seems to be the reasonable one.

But the truth is, neither theory is correct. Inflammation does not consist essentially in either an increased or decreased strength of action in the part inflamed; nor is there necessarily any augmentation or diminution of the vital energies of the part. Inflammation is simply *irregular* or *abnormal* action. Whether the action be strong or weak, so far as the circulation of the blood in the part is concerned, is quite immaterial. It may be one or the other in the first instance; but as the disease is prolonged, the blood-vessels soon become congested and over-distended; the accumulated blood soon distends their coats beyond the power of normal contraction, so that debility soon becomes the permanent condition.

But debility, or decreased action, is not to be remedied by irritating, exciting, and disturbing the vital energies with stimulants. To relieve the distended and weakened vessels, the destruction must be removed, and the part allowed to rest. The blood should be determined to other parts, not taken out of the body.

Nor is increased action to be "cured" by antiphlogistics. It may be "reduced" or *subdued*, and so may all vital action; but this is only a process of subduing the patient. A sick person does not possess too much blood, nor too much vital power. Sickness does not add to his capital stock of vitality; nor does a preternatural supply of vital energy ever occasion disease, for it never exists. The difficulty, in all cases of inflammation, and in all cases of disease, is in unbalanced determination of vital action, and in irregular distribution of the blood, as I have heretofore explained.

This subject has, perhaps, some special importance at this time, because the old and oft-exploded doctrines of a by-gone age—that bleeding and other reducing measures should be resorted to in treating diptheria, *because* it is an inflammatory affection—and that typhus fever is a *result* and not an *attendant* of inflammation, are being revived by modern physicians.

As an illustration of the propriety of these remarks, and as a basis of some further criticisms on the subject, the following article, which I find in one of the New York newspapers, of large circulation, is subjoined:

"AN INTERESTING MEDICAL PAPER.

"SORE THROAT, REMITTENT AND TYPHUS FEVERS.

"These diseases are to a considerable extent prevailing, and with some fatality. Any remarks about the treatment which tends to cure them will be of service to the people.

"Some time ago, a lecture was delivered at the Medical College in Twentieth Street, by Dr. Sherrill, on the epidemic sore-throat distemper. It was published in the *Christian Messenger*. In the remarks made, it was assumed and shown that the throat disease was purely of an inflammatory nature; that the symptoms of typhus, gangrene, or what is called diptheria, are effects of inflammation, and very likely may be avoided by active and suitable means early used to check such an inflammation. To effect this, all irritating stimulants and alcoholic mixtures are not advisable and are injurious. A great many facts and authorities are introduced to sustain this theory; it is stated that by the mode of treatment detailed in this essay, more than two hundred cases, in various states and stages, have been treated, and that with two exceptions all recovered.

"In illustration of this subject, it is stated that a typhous or gangrenous condition has been represented to have taken place in the throat disease, for which opium, stimulants, and alcoholic articles have been recommended and given. It appears that a state of typhus is preceded by fever of an inflammatory nature, as it is in this case, and also in remittent fever which precedes a typhoid state when it takes place, and that in the first stage it is of an inflammatory or congestive type. This is a very interesting position to take, and if it is correct, may or ought to have an important influence on the treatment. It may be a means of inducing prescribers to avoid the free use of stimulants and alcohol, which are frequently used. There are many authorities named in favor of these statements. The address will further explain on this subject by an extract:

"'It is doubtful whether in this climate any febrile disease, in its incipient stage, is of a typhous condition, so as to be benefited early by stimulant or alcoholic articles. The state of typhus which takes place according to the writer's observations has been the sequence of an inflammatory or congestive state of the body not arrested in the early stage.'

"Typhus was considered to be preceded by, and be the result of, an inflammation of some of the inner organs, by Clutterbuck, Armstrong, Broussais, Rush, Donaldson and Maygell, and that at first the treatment should be to relieve and cure such a condition of the body. Remittent fevers, which prevail in summer and autumn, in the first stage are of an inflammatory or congestive state, and a depleting and refrigerant course of treatment is always the most successful. This will be a means of checking the progress of the case and preventing a state of typhus from taking place, and the case may be cured in much less time than those cases are generally cured. A great many years of observation and practice in treating and curing many hundreds of such cases justifies these statements.

"The following sketches are taken from a collection of essays on epidemic diseases collected by Dr. H. Sherrill:

"In 1825 the remittent and typhus fever prevailed in many places along the Hudson River, in an epidemic form, and more than commonly severe. Some sketches of a history of it was read at the annual meeting of the Duchess County Society that fall. An abstract from it is here made: 'Generally the cases exhibited inflammatory action, but in many instances there was a small flaccid pulse, like that which often took place in the epidemic of 1812. There was dull or intense headache—dull appearance of the eyes—a lurid face—a tired aching of the limbs—the tongue was contracted, pointed, and very red, in some cases there was great prostration and congestion; when the disease was not checked early, a state of typhus set in, and this might be tedious, obstinate, or fatal.'

"The most suitable treatment was, in the early stage, to use active means to remove congestion and an inflammatory state; for this purpose free blood-letting was the most useful remedy; all the after-symptoms were shaped or controlled by the use or omission of this remedy. In those cases where there was great depression or congestion, and the pulse was small and flaccid, as it generally is in such a state of disease, the portion of blood taken at first was small, and the operation repeated, as was practiced in the epidemic of 1812 and in that of 1793, as recommended and used by Dr. Rush; in such a condition, the pulse always on bleeding rises and is more full and firm · the blood was very black; the medicine used was of a refrigerating, sudorific nature. There were as many as twenty-five cases treated in this way (which was large for a sparse-settled country district); the fever run out, and a crisis formed the ninth day; there were no stimulants given till after the crisis, and very little then; nourishment was mostly relied upon to restore the strength. Many of the cases assumed a typhous state, but it was soon controlled; there was not one case fatal.

"In many instances, and in most places, from information received and reports made, attempts were made to cure this disease by alexipharmic remedies, such as mercury, opium, sudorific cordial, and alcoholic mixtures, and those were freely used; in this way a long, tedious illness ensued; the case run on three, four, five, or six weeks; the patient got a black tongue and teeth, stupor, delirium, nervous irritation, and a train of those symptoms called typhous; frequently the case terminated in death.

"It is a fair inference to make, that under similar treatment corresponding results would occur at this time.

"Cases similar to those detailed, it appears, at this time may be cured or prove tedious and fatal according to the mode of treatment, as the following may show:

"'R. I., *Dec.* 20, 1861.

"'Took a cold, which increased so that on the 28th he took to bed with fever, pain in the head, nausea, soreness and aching of the limbs, rather prostrated pressure of the chest. Jan. 2, 1862: When I first saw him he was inclined to stupor—tongue contracted and red—bowels costive; he appeared to have foaming congestion of the brain or lung—pulse compressible, a beginning of the state called typhoid.

"'From the arm sixteen ounces of blood were taken; it was black, and deprived of a required quantity of vital air; it soon was as firm as liver; means were used to open the bowels; he was put upon the use of homeopathic medicine; these were varied from time to time to operate on the symptoms presented. After the use of the first remedies, the pulse became more full and firm. This was the case, by such means, in the epidemic of 1812, and in the epidemic cholera. He was given as much cold water as he would take; he took no nourishment, except gruel or the like; all stimulant and alcoholic articles were excluded; the fever and disease gradually abated. On the 10th the fever subsided and a crisis formed, so that on the 12th he set up, and for a short time read the news; the tongue retained a redness, attended with flushes of fever, so that remedies were still given to remove those symptoms of a trail of inflammatory action; even nourishment was sparingly given, and no alcoholic mixtures were allowed. He regularly improved, daily walked the room, and by the 20th went down stairs, fully cured. MEDICUS.'

"The following case took place about the same time as the preceding one. It was communicated by the nursery attendants:

"A. B. was attacked precisely like the other. After about ten days' lingering was taken to bed with symptoms similar to the preceding case. In four days he was greatly prostrated and distressed; the tongue was very red and pointed; it was now around that he had typhoid fever; before the fever ended or a crisis formed he was given beef-tea, and was soon put upon the use of port wine; he soon inclined to stupor and indifference; by the advice of several doctors, called respectable prescribers, the stimulants were increased; a black scurf formed on the tongue and teeth, the edges of the tongue retaining a lively redness. To keep him from running down and sinking, brandy was added to the other means; he was very uneasy, attended with nervous irritation and an impaired mind. In this way he struggled along for five weeks, and then died.

"In a history of the epidemic of 1812, as it appeared in Duchess County, which was described as a remitting bilious fever, in many cases it was attended with inflammation and congestion of some of the internal organs. When not early checked or relieved, it was strongly inclined to pass into a state of typhus or gangrene. In a township containing 2,400 inhabitants, there were about 130 cases. It was generally looked upon by the people and the medical men as a state of direst weakness and of a typhoid tendency, and recourse was had to a free use of a great variety of stimulants and alcoholic mixtures to keep off typhus, 'to keep the patient from running down into typhus and gangrene.' A clergyman of the place set down the names of those who died. It footed 63—one half!

"In another district, of about the same population, there were about 150 cases of the same epidemic. They were treated by free blood-letting and refrigerant remedies. No alcohol was used until the congestion and fever were removed and a crisis formed. Of these it is stated that ninety-four per cent. were cured.

"Several years afterward the work of Surgeon-General Mann and Prof. Gallup appeared, which treated on this epidemic of 1812. They advocated and recommended the same doctrine and practice which has just been mentioned, by which it appeared that in the army the proportion cured was over ninety per cent.

"In those essays of Messrs. Mann and Gallup it is stated that in the vicinity of the army, among the people, a stimulant treatment for the epidemic frequently was used, and that one fourth to one half the cases were fatal.

"In a report from Dr. Lovell, it is stated that in one village and vicinity where the stimulant practice was freely used, in the month of January, 1812, there were seventy-three deaths. During this time there were one hundred cases in the army, which were treated by blood-letting, etc., of which only three proved fatal.

"With such glaring statements before us, is it not surprising that a great and obstinate prejudice against bleeding exists in the community? By it, no doubt, many a one has lost his life. If the Sanitary Board of the army could be induced to review the facts, and they could be made to produce a fair influence on the general mind, many of the soldiers might have their lives preserved, who now fall victims to typhus, quinine, and alcohol, and the government might save the immense sums which these popular drugs cost."

It would be difficult to compress a greater number of pathological errors and therapeutic mistakes into so small a compass. Scientifically, it is a mere "budget of blunders." But as it represents the theory and practice of about an equal moiety of the medical profession of the whole civilized world, it is entitled to candid consideration and respectful refutation.

Dr. Sherrill assumes that typhus is an *effect* of inflammation, and that by promptly reducing the inflammation, that is, by employing bleeding and antiphlogistic drugs in the early stage, the consequential "typhous condition" may be averted.

Nothing can be further from the truth, and no doctrine could be more mischievous in practice. It has already slain its millions. Typhus or typhoid fever, or a "typhous condition," means nothing more nor less than a continued fever, with or without the concurrence of an acute local inflammation, in which the remedial effort is *not* chiefly and persistently determined to the whole surface of the body. If the determination of remedial effort—the "fever," or "reaction" of medical authors—is decidedly and permanently directed to the whole surface, the fever is properly called entonic or sthenic; otherwise it is asthenic, atonic, or typhous. In all forms of remittent fever, diptheria, malignant scarlet fever, putrid or epidemic sore throat, the diathesis, both of the local and the constitutional disease—the inflammation and the fever—is invariably atonic in all stages. The disease commences with the "typhous condition," progresses with the "typhous condition," and ends with the "typhous condition." And the effect of bleeding and other reducing processes and agents is always to aggravate the "typhous condition," and, when the patient is so lucky as to survive the disease and medication, to prolong the convalescence, and render recovery imperfect.

Dr. Sherrill does well in objecting to alcoholic and other stimulants, notwithstanding we have a formidable array of authorities in their favor; but he resorts to antiphlogistics *because* stimulants are injurious, he commits an error quite equal, and even more disastrous in results, to the mistake of those who recommend alcoholic stimulants *because* antiphlogistics are injurious.

Is there not something marvelously strange in this controversy? Was the like of it ever heard of or thought of on the face of the earth? Were scientific men ever in a similar muddle on any other subject?

Here is a learned body of men—40,000 strong in the United States—divided into two classes nearly equal in number, character, and experience—one class condemning stimulants and approving antiphlogistics, and the other class condemning antiphlogistics and approving stimulants. And each party refers to its own observations and experience to prove that its own practice is all right, and that the opposite treatment is all wrong. What can such experience be worth?

And so the profession might go on another three thousand years, ravaging the human constitution with poisonous drugs, and sending the human race in constant droves to premature graves, and justify their doings by "observation and experience." Never, never will their destroying hand be stayed until a true theory is understood, by which the facts of observation and experience may be judged and applied.

Whether a stimulant or an antiphlogistic plan of treatment is most successful, or rather, whether one or the other is less injurious, depends entirely on the degree of atony or debility—on the greater or less degree of the asthenic or typhoid condition of the system. In ordinary cases there would be little to choose. In these cases the patient will recover in spite of a great amount of injurious medication, whether of the stimulant or the antiphlogistic kind; and as the great majority of cases are of this character, the "success" of either plan, according to the ordinary operations of the law of chance, would be nearly equal. And this fact alone solves the problem why it is that the advocates of the opposite plans of treatment can never agree. But in the mildest cases the stimulant plan would be worse than the antiphlogistic; while in the severest cases the antiphlogistic treatment would be more fatal than the stimulant.

The subject is, however, still further complicated by the different degrees or potencies of stimulation or antiphlogistication to which different physicians resort. Some of the stimulating doctors use mild stimulants in moderate doses, while others employ strong stimulants in large doses. And so with the antiphlogisticating practitioners. Some employ the most deadly reducing agents of the materia medica, while others prescribe only the milder poisons of the same class. All of these circumstances must be taken into the account in estimating the effects of either plan of treatment. And, I apprehend, the perfect understanding of the whole subject will bring the reader to the conclusion that all physicians have come to, who have fully investigated the subject, viz.: the less drug-medication of any kind, the better for the patient.

Works by Dr. Trall.

Hydropathic Encyclopedia. A complete System of Hydropathy and Hygiene. With nearly one thousand pages. Fully illustrated...............	$3 00
Hydropathic Cook Book. With new Receipts.............................	0 87
Uterine Diseases and Displacements. With original engravings	2 50
Home Treatment for Sexual Abuses	0 30
The Alcoholic Controversy. Review of the *Westminster Review*.........	0 30
The Complete Gymnasium. Illustrated.	1 25
Tobacco: Its History, Nature, and Effects. A Prize Essay.	0 10
Philosophy of the Temperance Reformation. A Prize Essay	0 10
Diseases of the Throat and Lungs.....................................	0 20
Water-Cure for the Million	0 25
Lecture on Vegetarianism	0 10
"Nervous Debility." For Young Men	0 10
Lecture on Diseases of Females...............	0 10
Lecture on Drug Medicines	0 10
Sexual Diseases, including Venereal Affections	1 50

PATHOLOGY.

Diptheria. A complete and comprehensive work.	1 00
Principles of Hygeio-Therapy, and College Catalogue.............	0 10
The True Healing Art. His Washington Address......................	0 25
Anatomical and Physiological Plates, representing all of the important Structures and Organs of the Human Body, *in situ* and of the size of life. Price of the series of Six Plates, colored and mounted on rollers	12 00

WORKS IN PREPARATION BY DR. TRALL.

Physiology and Hygiene for Schools.......................	1 25
Sexual Physiology Complete	1 25

Principles of Hygienic Medication.—This work will embody the substance of the author's lectures to the Medical Classes of the Hygeio-Therapeutic College, with extensive statistical and scientific data to illustrate and confirm the principles it advocates and the problems it advances, compiled from the highest authorities of the different medical schools. It will contain a complete and thorough exposition of the Truths and Errors of all Medical Theories and Systems, with full and precise details and directions for the practical application of the Hygeio-Therapeutic system to the Home Treatment of all known diseases. It will also present, by way of contrast, a concise statement of the method of treating each disease, according to the doctrines of the different drug-medical schools. The work will consist of Three Volumes, of 750 pages each. Price $3 per Volume; to the Trade, $2. All single orders (with remittances) received in advance of the day of publication, will be supplied at the wholesale price—$6 for the Three Volumes. *Special terms to Traveling Agents.*

The Hygienic Teacher. This is a monthly publication of 24 quarto pages, edited by Dr. Trall, and published by Fowler and Wells, 308 Broadway, New York, devoted to the advocacy and explanation of the Hygienic Medical System. It is emphatically a health journal for the people. Price $1 a year.

NEW YORK HYGIENIC MEDICAL INSTITUTE.

Dr. Trall's institution for the reception and treatment of all classes of invalids is at No. 15 Laight Street, New York. It is the oldest institution of the kind in the country, and the largest city establishment in the United States. Dr. Trall is provided with competent male and female assistants for city and country practice.

HYGEIO-THERAPEUTIC COLLEGE.
R. T. TRALL, M.D., PRINCIPAL.

The regular lecture terms commence on the second Monday of November in each year, and continue twenty weeks. Particular attention is devoted to Practical Anatomy, Dissections, and Obstetrical Demonstrations. This school is chartered by the Legislature, and authorized to confer the degree of M.D. Fees for the whole course $75. Graduation fee $25.

NEW YORK
HYGEIO-THERAPEUTIC
COLLEGE.

[CHARTERED BY THE LEGISLATURE.]

The Course of Lectures for the Winter Term will commence the second Monday in November of each year, and continue twenty weeks, including one week's vacation during the holiday season.

FACULTY.

R. T. TRALL, M.D., INSTITUTES OF MEDICINE, THEORY AND PRACTICE, MATERIA MEDICA, FEMALE DISEASES, AND MEDICAL JURISPRUDENCE.
O. T. LINES, M.D., ANATOMY AND SURGERY.
HULDAH PAGE, M.D., PHYSIOLOGY AND HYGIENE.
A. K. EATON, M.D., CHEMISTRY AND NATURAL PHILOSOPHY.
LYDIA F. FOWLER, M.D., OBSTETRICS.
DR. H. F. BRIGGS, PHILOSOPHY OF VOICE, SPEECH, AND GESTURE.
L. N. FOWLER, A.M., PHRENOLOGY AND MENTAL SCIENCE.

CURATORS.

G. F. ADAMS, M.D., BROOKLYN, N. Y.
JAMES C. JACKSON, M.D., DANSVILLE, N. Y.
HULDAH PAGE, M.D., AUGUSTA, MAINE.
O. T. LINES, M.D., WILLIAMSBURGH, N. Y.
E. P. MILLER, M.D., NEW YORK.

Fees for the whole Course, $75, payable in advance. Matriculation Fee, $5. Graduation Fee, $20. Candidates for the degree of M.D. are required to deposit a thesis on some medical subject, with the Graduation Fee, two weeks before the close of the term.

Board can be had in the city for from $3 to $5 per week, according to rooms and other accommodations required. Students who prefer, can hire rooms and board themselves.

Programme of Educational Exercises.

Usually there will be Four Lectures daily, of one hour each. Half an hour, morning and evening, will be devoted to gymnastic and elocutionary exercises. A Clinique will be held every Friday; and on Saturdays the students will visit the hospitals and public institutions, where a great variety of surgical operations are performed, and where almost every phase of diseased and deformed humanity can be seen. There will be a Lyceum debate on Medical subjects one or two evenings of each week, with criticisms, essays, etc., by members of the class, for mutual improvement. One or two evenings of each week will be appropriated to music, dancing, and other wholesome exercises and recreations.

For further information, address

R. T. TRALL, M.D.,
15 LAIGHT STREET, NEW YORK.

"BOOKS SENT, PREPAID, BY MAIL, TO ANY POST-OFFICE IN THE UNITED STATES."

A LIST OF WORKS

BY

FOWLER AND WELLS, 308 BROADWAY, NEW YORK.

In order to accommodate "the people" residing in all parts of the United States, the Publishers will forward, by return of the first mail, any book named in the following list. The postage will be pre-paid at the New York office. The price of each work, including postage, given, so that the exact amount may be remitted. Letters containing orders should be post-paid, and directed as follows: FOWLER AND WELLS,
308 BROADWAY, NEW YORK.

WORKS ON PHRENOLOGY.

COMBE'S LECTURES ON PHRENOLOGY. A complete course. Bound in muslin, $1 25.

CHART for Recording various Developments. Designed for Phrenologists. 6 cents.

CONSTITUTION OF MAN. By Geo. Combe. Authorized edition, with Illustrations, embracing his Portrait. Muslin, 87 cents.

DEFENCE OF PHRENOLOGY, Arguments and Testimony. By Dr. Boardman. A work for Doubters. Muslin, 87 cents.

DOMESTIC LIFE, THOUGHTS ON; its Concord and Discord. By N. Sizer. 15 cents.

EDUCATION COMPLETE. Embracing Physiology, Animal and Mental, Self-Culture, and Memory. One large vol. By Fowler. $2 50.

EDUCATION, founded on the Nature of Man. By Dr. Spurzheim. Muslin, 87 cents.

FAMILIAR LESSONS ON PHRENOLOGY and PHYSIOLOGY. An excellent work for Children. Beautifully Illustrated. $1 25.

MARRIAGE; its History and Philosophy, with directions for Happy Marriages. Bound in muslin, 74 cents.

MATRIMONY; or, Phrenology and Physiology applied to the Selection of Congenial Companions for Life. By Fowler. 30 cents.

MEMORY AND INTELLECTUAL IMPROVEMENT; applied to Self-Education. By Fowler. Muslin, 87 cents.

COMBE'S MORAL PHILOSOPHY; or, The Duties of Man. By Geo. Combe. 87 cents.

MENTAL SCIENCE, Lectures on, according to the Philosophy of Phrenology. By Rev. G. S. Weaver. Muslin, 87 cents.

PHRENOLOGY PROVED, Illustrated, and Applied. Thirty-seventh edition. A standard work on the science. Muslin, $1 25.

PHRENOLOGICAL JOURNAL, American, Monthly. Quarto, Illustrated. A year, $1.

PHRENOLOGY AND THE SCRIPTURES. By Rev. John Pierpont. 15 cents.

PHRENOLOGICAL GUIDE. Designed for the Use of Students. 15 cents.

PHRENOLOGICAL ALMANAC. Illustrated with numerous engravings. 6 cents.

PHRENOLOGICAL BUST: designed especially for Learners, showing the exact location of all the Organs of the Brain fully developed. Price, including box for packing, $1 25. (Not mailable.)

PHRENOLOGICAL SPECIMENS for Societies and Private Cabinets. 40 casts; nett, $25.

SELF-CULTURE AND PERFECTION OF CHARACTER. Muslin, 87 cents.

SELF-INSTRUCTOR in Phrenology and Physiology. Illustrated with one hundred engravings. Muslin, 50 cents.

SYMBOLICAL HEAD and Phrenological Chart, in map form, showing the Natural Language of the Phrenological Organs. 30 cents.

COMPLETE WORKS OF DR. GALL on PHRENOLOGY. 6 vols., $7.

THE PUBLISHERS would respectfully refer Strangers, Agents, and Country Dealers to the principal Publishers in New York, Philadelphia, Boston, or other cities, for evidence of their ability to fulfill all contracts. They have been many years before the public, engaged in the publishing business in the city of New York.

FOWLER AND WELLS'S PUBLICATIONS.

HYDROPATHY; OR, WATER-CURE.

IF THE PEOPLE can be thoroughly indoctrinated in the general principles of HYDROPATHY, and make themselves acquainted with the LAWS OF LIFE AND HEALTH, they will well-nigh emancipate themselves from a need of doctors of any sort.—DR. TRALL.

By no other way can men approach nearer to the gods than by conferring health on men.—CICERO.

ACCIDENTS AND EMERGENCIES. By Alfred Smee. Notes by Trall. Illustrated. 15 cents.

CHILDREN; their Hydropathic Management in Health and Disease. Dr. Shew. $1 25.

CHOLERA; its Causes, Prevention, and Cure, and all other Bowel Complaints. 30 c.

CONSUMPTION; its Causes, Prevention and Cure. Muslin, 87 cents.

COOK BOOK, HYDROPATHIC. With New Recipes. Illustrated. By R. T. Trall, M.D. Muslin, 87 cents.

DOMESTIC PRACTICE OF HYDROPATHY, with 15 engraved illustrations of important subjects from drawings. By E. Johnson, M.D. $1 50.

FAMILY PHYSICIAN, Hydropathic. By Dr. Shew. A new and invaluable work for home practice. Profusely illustrated. $2 50.

HYDROPATHIC ENCYCLOPEDIA: Illustrated. A Complete System of Hydropathy and Hygiene, embracing Anatomy, Illustrated; Physiology of the Human Body; Hygienic Agencies, and the Preservation of Health; Dietetics and Cookery; Theory and Practice of Treatment; Special Pathology and Hydro-Therapeutics, including the Nature, Causes, Symptoms, and Treatment of all known Diseases; Application to Surgical Diseases and to Hydropathy, to Midwifery and the Nursery. With Three Hundred Engravings, and nearly One Thousand Pages, including a Glossary, Table of Contents, and Index, complete. By R. T. Trall, M.D. Price, $3.

HYDROPATHY; or, Water-Cure Principles and Modes of Treatment. Dr. Shew. $1 2

INTRODUCTION TO THE WATER CURE. With First Principles. 15 cents.

PHILOSOPHY OF WATER-CURE By J. Balbirnie, M.D. A work for beginners. 30 c

PRACTICE OF WATER-CURE. B Drs. Wilson and Gully. A handy, popular work. 30

RESULTS OF HYDROPATHY treating of Constipation and Indigestion. By E ward Johnson, M.D. 87 cents.

WATER-CURE IN CHRONIC DIS EASES; an exposition of the Causes, Progress, and T minations of Various Chronic Diseases. By Dr. J. Gully. An important work. $1 50.

WATER AND VEGETABLE DIE in Scrofula, Cancer, Asthma, etc. By Dr. Lamb. Note by Dr. Shew. Muslin, 87 cents.

WATER-CURE IN EVERY KNOWN Disease. By J. H. Rausse. Muslin, 87 cts.

WATER-CURE MANUAL. A popular work on Hydropathy. Muslin, 87 cts.

WATER-CURE FOR THE MIL LION. 20 cents.

DISEASES OF THE THROAT AND LUNGS, including Diptheria. By Dr. Trall. 15 cents

ANATOMICAL AND PHYSIOLOGICAL PLATES. These Plates were arranged expressly for Lecturers on Health, Physiology, etc. By Dr. R. T. Trall, M.D., of the New York Hydropathic College. They are six in number, representing the normal position and life-size of all the internal viscera, magnified illustrations of the organs of the special senses, and a view of the principal nerves, arteri veins, muscles, etc. For popular instruction, for families, schools, and for professional reference, they will found far superior to anything of the kind heretofore published, as they are more complete and perfect in istic design and finish. Price for the set, fully colored, backed and mounted on rollers, $12. (Not mailable.

WATER-CURE JOURNAL AND HERALD OF REFORMS. Devote to Hydropathy, its Philosophy and Practice; to Physiology and Anatomy, with illustrative engravings Dietetics, Exercise, Clothing, Occupations, Amusements, and those Laws which govern Life and Health. Published Monthly, at $1 a year in advance.

SPECIAL LIST. We have, in addition to the above, Medica Works and Treatises on subjects which, although not adapted to general circulation, are invaluable to those w need them. This Special List will be sent on application.

FOWLER AND WELLS have all works on HYDROPATHY, PHYSIOLOGY, and the Natural Sciences generally. Boo sellers supplied on the most liberal terms. Agents wanted in every State, County, and Town. These works universally popular, and thousands might be sold where they have never yet been introduced. Orders shoul post-paid, and directed to the Publishers, as follows: **FOWLER AND WELLS,**

[Name the Post-office, County, and State.] 308 BROADWAY, NEW YORK

FOWLER AND WELLS'S PUBLICATIONS.

PHYSIOLOGY—MESMERISM—PHONOGRAPHY.

ON PHYSIOLOGY.

ALCOHOL AND THE CONSTITUTION OF MAN. Illustrated. By Youmans. 30 cents.

ALCOHOLIC CONTROVERSY. A Review of the *Westminster Review* on the Physiological Errors of Teetotalism. By Dr. Trall. 30 cents.

COMBE'S PHYSIOLOGY, applied to the Improvement of Mental and Physical Education. Notes by Fowler. Muslin, 87 cents.

DIGESTION, PHYSIOLOGY OF. The Principles of Dietetics. By Andrew Combe. 30 cts.

FAMILY GYMNASIUM. With numerous illustrations; Containing the most improved methods of applying Gymnastic, Calisthenic, Kinesipathic, and Vocal exercises to the development of the bodily organs, the invigoration of their functions, the preservation of health, and cure of diseases and deformities. By R. T. Trall, M.D. $1 25.

FAMILY DENTIST; a Popular Treatise on the Teeth. By D. C. Warner, M.D. 87 cts.

FOOD AND DIET; containing an Analysis of every kind of Food and Drink. By Dr. J. Pereira. Muslin, $1 25.

FRUITS AND FARINACEA the Proper Food of Man. With Notes and engraved Illustrations. By R. T. Trall, M.D. Muslin, $1 25.

HUMAN VOICE; its Right Management in Speaking and Reading. 25 cents.

INFANCY; or, the Physiological and Moral Management of Children. Illustrated. By Dr. Combe. Muslin, 87 cents.

NATURAL LAWS OF MAN. By Dr. Spurzheim. A good work. 30 cents.

PHILOSOPHY OF SACRED HISTORY, considered in Relation to Human Aliment and the Wines of Scripture. By Sylvester Graham. $2.

PHYSIOLOGY, Animal and Mental, applied to Health of Body and Power of Mind. By O. S. Fowler. Muslin, 87 cents.

SOBER AND TEMPERATE LIFE; with Notes and Illustrations by Louis Cornaro. 30 cts.

THE SCIENCE OF HUMAN LIFE. By Sylvester Graham, M.D. With a Portrait and Biographical Sketch of the Author. $2 50.

TEA AND COFFEE; their Physical, Intellectual, and Moral Effects. By Alcott. 15 cts.

TEETH; their Structure, Disease, and Management, with Engravings. 15 cents.

TOBACCO, Works on: comprising Essays by Trall, Shew, Alcott, Baldwin, Burdell, Fowler, Greeley, and others. Complete in 1 vol. 82 c.

VEGETABLE DIET, as sanctioned by Medical Men and Experience in all Ages. By Dr. Alcott. Muslin, 87 cents.

MESMERISM—PSYCHOLOGY.

ELECTRICAL PSYCHOLOGY, Philosophy of, in Twelve Lectures. By Dr. J. B. Dods. Muslin, 87 cents.

FASCINATION; or, the Philosophy of Charming (Magnetism). Illustrating the Principles of Life. Muslin, 87 cents.

LIBRARY OF MESMERISM AND PSYCHOLOGY. With suitable illustrations. In two large volumes of about 900 pages. Price, $3.

MACROCOSM AND MICROCOSM; or, the Universe Without and the Universe Within. By Fishbough. Scientific Work. Muslin, 87 cts.

PHILOSOPHY OF MESMERISM AND CLAIRVOYANCE. Six Lectures. With Instructions. 30 c.

PSYCHOLOGY; or, the Science of the Soul. By Haddock. Illustrated. 30 cents.

ON PHONOGRAPHY.

THE PHONOGRAPHIC INSTRUCTOR. By Ben Pitman. Elementary. 30 cents.

THE MANUAL OF PHONOGRAPHY. By Pitman. A new and comprehensive exposition of Phonography, with copious illustrations and exercises. Useful for beginners. 60 cents.

AMERICAN MANUAL OF PHONOGRAPHY; being a complete guide to the acquisition of Short-Hand. By Elias Longley. 60 cents.

PHONOGRAPHIC TEACHER; being an Inductive Exposition of Phonography, with instructions to those who have not the assistance of an Oral Teacher. By E. Webster. 45 cents.

PHONOGRAPHIC COPY-BOOKS, with Morocco Covers, for the use of students. 50 cts.

All works on Phonography will be furnished to order, by FOWLER AND WELLS, 308 Broadway, New York.

Either of these works may be ordered and received by return of the FIRST MAIL, postage prepaid by the Publishers. Please address all letters, post-paid, to

[Name the Post-office, County, and State.]

FOWLER AND WELLS,
308 BROADWAY, NEW YORK.

FOWLER AND WELLS'S PUBLICATIONS.

MISCELLANEOUS AND HAN[D

When single copies of these works are wanted, the amount, in postage stamps, [...] may be inclosed in a letter and sent to the Publisher, who will forward the books by [...]

MISCELLANEOUS.

AIMS AND AIDS FOR GIRLS AND YOUNG WOMEN. By Rev. G. S. Weaver. Price, 87 c[...]

CHEMISTRY, applied to Physi-ology, Agriculture, and Commerce. By Liebig. 25 cts.

DEMANDS OF THE AGE ON COL-LEGES. An Oration. By Horace Mann. 25 cents.

DELIA'S DOCTORS; or, a Glance Behind the Scenes. By Miss Hannah Gardner Creamer. For the Family. Muslin, 87 cents.

FRUIT CULTURE FOR THE MILL-ION; or, Hand-Book for the Cultivation and Management of Fruit Trees. Illustrated with Ninety Engravings. By Thomas Gregg. Muslin, 50 cts.

HINTS TOWARD REFORMS, in Lectures, Addresses, and other Writings. By H. Greeley. Second edition, with Crystal Palace. $1 25.

HOME FOR ALL; the Gravel Wall, a New, Cheap, and Superior Mode of Building, with Engravings, Plans, Views, etc. 87 cents.

HOPES AND HELPS FOR THE YOUNG OF BOTH SEXES. By Rev. G. S. Weaver. An excellent work. Muslin, 87 cents.

IMMORTALITY TRIUMPHANT.—The Existence of a God, with the Evidence. By Rev. J. B. Dods. Muslin, 87 cents.

KANSAS REGION; Embracing Descriptions of Scenery, Climate, Productions, Soil, and Resources of the Territory. Interspersed with Incidents of Travel. By Max Greene. 40 cents.

LECTURES ON THE SCIENCE OF HUMAN LIFE. By Sylvester Graham, M.D. $2 50.

DIPTHERIA, its Nature, History, Causes, Prevention and Treatment on Hygienic principles; with a *Resume* of the various theories and practices of the medical profession. By Dr. R. T. Trall, just published. Muslin, $1.25.

MOVEMENT-CURE. Embracing the History and Philosophy of this System of Medical Treatment. Fully illustrated. By Geo. H. Taylor, M.D. $1 25.

PHYSICAL PERFECTION; or, the Philosophy of Human Beauty; showing how to Acquire and Retain Bodily Symmetry, Health, and Vigor; Secure Long Life; and Avoid the Infirmities and Deformities of Age. An excellent work. $1.

HOW TO GET A PATENT, with Instructions to Inventors. 6 cents.

POPULATION, T[...] Law of Animal Fertility.

SAVING AND W[...] mestic Economy Illustrat[...]

THE RIGHT WO[...] PLACE: A Pocket Diction[...] Terms, Abbreviations, F[...]

WAYS OF LIFE; and the Wrong Way. B[...] ital work. Muslin, 60 c[...]

NEW HA[

HOW TO WRI[...] Manual of Composition [...] nable to the young. Pape[...]

HOW TO TAL[...] Manual of Conversation [...] Five Hundred Common [...] rected. Paper, 30 cents

HOW TO BEHA[...] Maunal of Republican Eti[...] Personal Habits, with [...] and Deliberative Assemb[...]

How to Do Bu[...] et Manual of Practical A[...] in Life, with a Collectio[...] Forms. Suitable for all.

HAND-BOOKS [...] PROVEMENT (Education[...] Write," "How to Talk[...] "How to Do Business,"

RURAL H[A

DOMESTIC ANI[...] ual of Cattle, Sheep, and [...] to Breed, Rear, and Man[...] yard. Paper, 30 cents;

THE FARM: Practical Agriculture; [...] Field Crops, with a mo[...] Management. Paper, 3[...]

THE GARDEN: Horticulture; or, How to [...] and Flowers. Paper, 30 [...]

THE HOUSE: Rural Architecture; or [...] Barns, and Out-Houses [...]

RURAL MANUA[...] "The House," "The [...] "Domestic Animals," [...]

These works may be ordered in large or small quantities. A liberal discount w[...] others, who buy to sell again. They may be sent by Express or as Freight, by R[...] place in the United States, the Canadas, Europe, or elsewhere. Checks or drafts, [...] York, Philadelphia, or Boston, always preferred. We pay cost of Exchange. All le[...] addressed as follows:

FOWLER AND [W...]

[Name the Post-office, County, and State.]